HEROES
&
SCOUNDRELS

THE GOOD, THE BAD, AND THE UGLY
OF THE
NOBEL PRIZE IN MEDICINE

Moira Dolan, MD

Quill
Driver
Books

Fresno, California

Heroes & Scoundrels: The Good, the Bad, and the Ugly
of the Nobel Prize in Medicine
© copyright 2022 Moira Dolan, MD
Cover image courtesy Shutterstock/Jolygon

ISBN: 978-161035-393-9
1 3 5 7 9 8 6 4 2

Linden Publishing titles may be purchased in quantity at special discounts for educational, business, or promotional use. To inquire about discount pricing, please refer to the contact information below.

For permission to use any portion of this book for academic purposes, please contact the Copyright Clearance Center at www.copyright.com

Printed in the United States of America
Library of Congress Cataloging-in-Publication data on file

Linden Publishing, Inc.
2006 S. Mary
Fresno, CA 93721
www.lindenpub.com

Contents

Preface...v

Chapter 1: Yellow Jack 1
Chapter 2: Dirty Business............................ 14
Chapter 3: Squiggles and Gold 22
Chapter 4: The Polio Researchers 29
Chapter 5: Like the Priest at a Wedding.............. 41
Chapter 6: Self-Experimentation...................... 47
Chapter 7: Drugs..................................... 60
Chapter 8: From Wahoo to Outer Space 67
Chapter 9: The Men (and Women) of DNA 77
Chapter 10: Self and Nonself......................... 87
Chapter 11: Postal Worker Wins the Nobel Prize....... 97
Chapter 12. The Wonder Boys.......................... 105
Chapter 13: What Nerve............................... 120
Chapter 14: The Cholesterol Discoveries.............. 133
Chapter 15: The French Freedom Fighters 142
Chapter 16: The Cancer Detectives.................... 153
Chapter 17: The Visionaries.......................... 166
Chapter 18: Making Protein........................... 178
Chapter 19: Viruses Everywhere....................... 187
Chapter 20: Brain Chemicals.......................... 202
Chapter 21: The Midwestern Genius 216
Chapter 22: Untangling Antibodies 221
Chapter 23: The Birds and the Bees 227
Chapter 24: Inside the Cell.......................... 249
Chapter 25: Infected by Viruses 265

Afterword.. 288
Winners of the Nobel Prize in Physiology or Medicine
1951–1975 291
Index 294

Preface

Welcome to the continuation of the lively biographies of winners of the Nobel Prize in Medicine begun in my previous book, *Boneheads and Brainiacs*, which covered the years 1901 to 1950. This volume profiles the prizewinners from 1951 to 1975. As a result, my research was easier, because so much more documentation is available from this period, including many video interviews; more numerous memoirs, biographies, and autobiographies; and treasure troves of the entire collected works of many of these scientists. Thus, the stories of only a quarter century fill as many pages as the first fifty years in *Boneheads and Brainiacs*.

Another benefit of the abundant documentation was that I was able to discover more about the sidelined players—especially the unrecognized women who conducted much of the prizewinning research themselves or alongside the men who would become Nobel Prize winners. Medical history buffs may have already heard of Rosalind Franklin's role in the discovery of the structure of DNA, but in these pages you will also meet Filomena Nitti, Esther Zimmer, Marianne Grunberg-Manago, Elizabeth Keller, Martha Chase, Ruth Hubbard, Betty Press, and Marguerite Vogt. In the first fifty years of the prize, Gerty Cori was the only woman to win, when she shared the award in 1947 with her husband, Carl Cori. In the next twenty-five years, all of the winners were male and white with the exception of Har Gobind Khorana, the only medicine prizewinner so far to come from India, also male.

Some themes are carried over from the first half century of the prize, such as the influences of the two world wars. In these pages, you will meet two winners who were card-carrying members of the Nazi Party, one famous American racist, a host of scientists who escaped the war as academic and political refugees, and amazing scientists who were resistance fighters. Other carry-overs from the first book are episodes of unethical behavior—notably taking credit where none is due—and, on the other end of the spectrum, instances of scientists not taking responsibility for goofs or fraud committed by others in their labs.

While some of the prizewinning research was truly delightful, like the discoveries of the amazing significance of the dance of the honeybees, other research was at best unoriginal, even leading a couple of these winners themselves to wonder why they got the prize. These are instances where advances in techniques led to mundane research yielding results largely due to nothing more than the application of good lab technique in a workmanlike fashion rather than any brilliant insight or novel approach to a scientific problem. Even the most famous of these accomplishments, the discovery of the structure of DNA, would have been worked out eventually by other researchers sooner or later, probably within weeks to months—it's just that Watson, Crick, and Wilkins beat everybody else to it.

The period covered here saw a shift in the nature of the scientific works that were recognized by a Nobel Prize. Discoveries in the first half of the twentieth century—such as penicillin, vitamin C, and estrogen—were more obviously physical and usually more directly applicable to patient care. In the next quarter-century, research largely turned toward entities visualized only with the aid of an electron microscope, or, more commonly, only indirectly identified and deduced through biochemical reactions. This research focused on genetics and viruses above all else. It is often difficult to discern the applicability of many of these discoveries to the everyday life of the health care consumer, but it does seem that the current pandemic has focused attention on these topics.

It is my hope that my readers become interested in the science and are entertained by the human stories.

Enjoy!
Moira Dolan, MD
Austin, Texas, May 2022

1

Yellow Jack

The Nobel Prize in 1951 was awarded to Max Theiler for his discoveries concerning yellow fever and how to combat it. The story of the research into yellow fever is the Nobel Prize's most deadly tale. It started more than half a century before the prize was awarded, and it is strewn with the illnesses and deaths of many researchers along the way. While it was not unusual for early infectious-disease researchers to fall victim to the illnesses they studied, yellow fever caused more sickness and death in investigators than any other disease.

The yellow fever victim suddenly feels feverish and becomes agitated or irritable. They then get a headache that becomes piercing in intensity and is accompanied by extreme light sensitivity Within hours, the victim's temperature goes up to 103 degrees Fahrenheit or higher. At the same time, their pulse slows down, which prevents sweating, and as a result, the victim rapidly becomes dehydrated. The disease next attacks the internal organs, including the kidneys, intestines, liver, and brain. Liver failure causes a buildup of yellow bile, resulting in the skin turning the yellow color of saffroned rice and the whites of the eyes taking on a golden glow—the intense coloration yellow fever is named for. At this point, some patients may begin to slowly recover. The unlucky progress to vomiting black blood and may even bleed from every orifice. The most freakish aspect of yellow fever is how it can affect the brain, causing agonizing delirium and violent convulsions until death. Complete recovery can take weeks or months, and even then, in rare cases a person can die from heart complications years after apparent recovery. Modern medical literature reports yellow fever morality worldwide at over 40 percent.

The infectious disease originated in Africa, where it was endemic—present all the time at low levels. Widely fatal epidemics of the disease at higher levels in specific regions were not recorded until an outbreak in 1648 in Barbados in the eastern Caribbean. More outbreaks followed the next year in Mexico's Yucatán and in Brazil, after ports in both places received slave ships. The United States saw a yellow fever epidemic the following year in New York, again linked to the arrival of a slave ship. Subsequently, there were epidemics in Philadelphia, where in 1793 some 9 percent of the population was killed; Baltimore; and again in New York City. The 1800s saw major epidemics in Charleston, Savannah, Mobile, New Orleans, and Memphis. Memphis was hit a second time in 1878 with an outbreak more deadly than ever before experienced in the US. The first two cases were recorded at the end of July. By August, yellow fever deaths were so numerous that there was a mass exodus from the city. By September there were only 19,000 residents left, of whom an estimated 17,000 were infected. Ultimately there were over 5,000 deaths. This incident remains the largest and most deadly urban infectious epidemic to hit America, relative to a city's population.[1]

Yellow Fever Compared to Coronavirus

Yellow fever and COVID-19 are both caused by viruses. In Memphis in 1878, yellow fever claimed about 5,000 lives out of a total population of approximately 33,000. In 2020, there were 891 deaths attributed to COVID in Shelby County, which includes Memphis, out of a population of about 937,000.

1   K. D. Patterson, "Yellow Fever Epidemics and Mortality in the United States, 1693–1905," *Social Science & Medicine* 34, no. 8 (1992): 855–65.

Ships with the yellow fever on board were quarantined offshore because it was believed that the infected sailors could spread the disease. This gave rise to the disease being called yellow jack, the same name given to the bright yellow cautionary flag that quarantined ships returning from the tropics were once required to hoist as they waited beyond the harbor until there were no more signs of fever in their crew. When yellow fever brought down several residents of a dockside neighborhood in Memphis, the homes of infected people were boarded up with the victims inside in what turned out to be a futile effort to contain the pestilence. After the occupants were dead, only special body handlers were allowed to touch the remains and the clothing was burned. As the death toll mounted, the streets were sloshed with a foul-smelling slurry of ferrous sulfate and carbolic acid to kill unspecified germs. The post office punctured holes through the stamps of outgoing letters to fumigate them with antibacterial sulfur. These precautions made some sense at the time, when the cause of yellow fever remained entirely unknown and germ theory had recently come into vogue, especially with the discoveries by Louis Pasteur and Robert Koch.

––––––––

By the late 1800s only humans had been known to get yellow fever, and the lack of an animal model in which to study the infection greatly slowed research. As early as 1881, the Cuban researcher Dr. Carlos Finlay proposed that the infectious particle of yellow fever was transmitted by mosquitoes. Finlay attended the Medical College of Pennsylvania then obtained more training in France before returning to Cuba. In August of 1881 Finlay presented his idea to an audience of scientists and doctors at the Royal Academy of Medical, Physical and Natural Sciences in Havana by reading his paper "The Mosquito Hypothetically Considered as the Agent of Transmission of Yellow

Fever."[2] He was ridiculed in scientific circles for this opinion, yet it was not a ridiculous idea. Malaria was an unrelated tropical disease that also happened to cause fevers and liver failure leading to yellow jaundice. The Scottish tropical-disease expert Dr. Patrick Manson had suggested that malaria was transmitted by mosquitoes as early as 1877, and it was finally proven by Charles Laveran in 1884, followed by Ronald Ross's documentation in 1897 of the complete life cycle of a malaria parasite in the mosquito. Ross and Laveran were winners of the Nobel Prizes in 1902 and 1907, respectively.

As with so much Nobel Prize–winning research of the era, the necessity to solve the riddle of yellow fever was driven by commerce and war. In the 1880s, the US was a major consumer of Cuban exports, including sugar, tobacco, cacao, coffee, tropical fruits, and nuts, which arrived exclusively by ship. Yellow fever was endemic in Cuba and episodically occurred in US harbors that received Cuban ships. When the US took Cuba in the short Spanish-American War of 1898, they lost only four hundred soldiers in combat, but yellow fever afflicted more than two thousand troops. This led the American occupiers to send a commission to Cuba from Washington headed by Walter Reed, research physician and major in the US Army.

The Havana Yellow Fever Commission decided to proceed with human experimentation, a grave undertaking in light of the high chances of a gruesome illness and horrible death. Volunteers were recruited by promising one hundred dollars to US Army personnel stationed in Cuba who were not enlisted soldiers and by luring Spanish immigrants with one hundred dollars in gold to sign up and the promise of an additional hundred-dollar bonus if they came down with yellow fever. (Today, each hundred-dollar payment would be worth around $3,000.) Informed consent was insisted upon by the American

---

2   C. J. Finlay, "El mosquito hipotéticamente considerado como agente de transmisión de la fiebre amarilla," *Anales de la Real Academia de Ciencias Médicas Físicas y Naturales de la Habana* 18 (1881): 147–69, reprinted in *Medical Classics* 2, no. 6 (1938): 590.

governor-general of Cuba, Brigadier General Leonard Wood. The document signed by volunteers read in part: "The undersigned understands perfectly well that in the case of the development of yellow fever in him, that he endangers his life to a certain extent but it being entirely impossible for him to avoid the infection during his stay in the island, he prefers to take the chance of contracting it intentionally in the belief that he will receive from the said Commission the greatest care and the most skillful medical service."[3] There were only a handful of volunteers. Ultimately, several of the commission's own researchers offered themselves as test subjects, perhaps more terrifying for them than the ordinary person, since they had seen the horrors of yellow fever deaths up close.

In one experiment, a small structure called the Infected Clothing Building was erected to test if human waste or soiled clothing would transmit the disease. The commission's own researchers served as volunteers, living in the house for twenty days wearing clothing—often caked with dried blood and feces—taken from recently deceased yellow fever victims. No one came down with the fever, although the volunteers almost went mad with isolation, fear, and disgust. Walter Reed and his team concluded that yellow fever was not transmitted by clothing or poor sanitation.

Walter Reed then started to warm to the idea of mosquito transmission, but it needed to be proven with more human experimentation. The research team kept containers of mosquitoes and allowed them to "feed" on yellow fever victims, then these same mosquitoes were allowed to bite a volunteer. Among others of the Havana commission, Dr. James Carroll let a sickly mosquito bite him. The mosquito perked right up after his blood meal, and Carroll soon came down with a fever. Before he became too ill to work, Carroll pulled a sample of his own

---

3   Contract between Antonio Benigno and Yellow Fever Board (English translation), November 26, 1900, Philip S. Hench Walter Reed Yellow Fever Collection, Box 70, Folder 4, Document 1, University of Virginia Claude Moore Health Sciences Library, Charlottesville, VA, reprinted in *Military Medicine* 166, Suppl I (2001): 38–9.

blood and put it through a Berkefeld filter, which uses diatomaceous material—the skeletons of one-celled algae—with a pore size so small that it traps bacteria but allows viruses slip through. Whatever remains after passing through a Berkefeld filter is a virus. Carroll found that the infectious agent in his blood was filterable, and thus that the responsible germ was a virus. We now know that yellow fever viral particles are about fifty nanometers in width, which is forty times smaller than the largest bacteria.

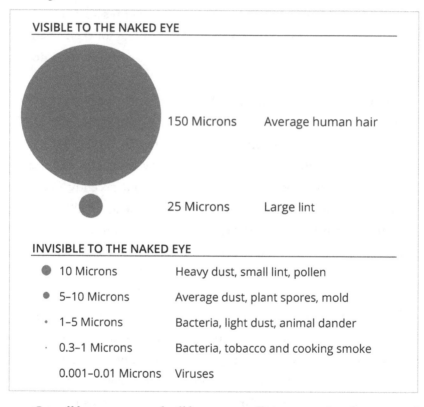

VISIBLE TO THE NAKED EYE

150 Microns	Average human hair
25 Microns	Large lint

INVISIBLE TO THE NAKED EYE

10 Microns	Heavy dust, small lint, pollen
5–10 Microns	Average dust, plant spores, mold
1–5 Microns	Bacteria, light dust, animal dander
0.3–1 Microns	Bacteria, tobacco and cooking smoke
0.001–0.01 Microns	Viruses

Carroll became severely ill but eventually recovered and continued his work with the project. This was followed by more human experimentation, including by thirty-four-year-old commission member Dr. Jesse Lazear, who entered himself in his own laboratory logbook as "Guinea Pig #1" when he let a research mosquito bite him. Lazear died of yellow fever after a short illness in which he became psychotic and convulsive from the disease. Another volunteer was Clara Louise

Maass, an army contract nurse from New Jersey. She allowed herself to be bitten by yellow fever–infected mosquitoes on five different days but developed no signs of infection. In another round of experimentation, she was bitten by four infected mosquitoes on the same day. Within four days Maass became seriously ill, and within a week, she was dead at the age of twenty-five.[4] Her death sparked strong public objection, including this statement by her mother as quoted in a New Jersey newspaper: "From what I know of the circumstances, my daughter's death seems little short of murder."[5] In all, the Havana experiments involved fifty-four subjects with four deaths, including Lazear, Maass, and two Spanish immigrants, Antonio Carro and Cumpersino Campa. A fifth death was later added when Carroll died in 1907 of heart valve failure, ultimately determined to be a late complication of yellow fever.[6] The commission thus substantiated the mosquito-transmission theory of Dr. Finlay and also proved that the infecting agent was filterable (a virus). The cycle of the virus was soon determined: A mosquito carries the virus. When it bites a human, some of the mosquito's blood gets into a victim, transmitting the virus. The infected human gets ill as the virus multiplies in their blood. The infected human gets bitten again, and a second mosquito picks up fresh virus, which it can spread to the next human it bites.

The Nobel Committee received seven nominations for the role played by Major Walter Reed in his leadership of the Havana Yellow Fever Commission, but he did not outlast deliberations and the prize cannot be awarded posthumously. Reed had suffered with long-term digestive problems and died at his home in Pennsylvania in 1902, at the age of forty-nine, after surgery for a ruptured bowel, possibly due

---

4   E. Chaves-Carballo, "Clara Maass, Yellow Fever and Human Experimentation," *Military Medicine* 178, no. 5 (2013): 557–562.

5   R. D. Paine, "A Martyr to Science: Miss Clara Maas[s], of East Orange, Falls a Voluntary Prey to the Yellow Fever Mosquito," *Newark Sunday News*, September 1, 1901.

6   American Association for the Advancement of Science, "AAAS Resolution: Death of Dr. James Carroll from Yellow Fever Experimentation" (professional society resolution, December 30, 1907).

to appendicitis. Dr. Carlos Finlay also had seven nominations for the Nobel Prize but did not win by the time he died in 1915.

The yellow fever saga continued with Hideyo Noguchi, a Japanese-born researcher working at the Rockefeller Institute, who ignored the information that the yellow fever particle was filterable and thus the conclusion that it could not possibly be a bacterium. He researched yellow fever in Ecuador and claimed to discover that the responsible germ was a spiraled bacterium.[7] The Rockefeller Institute quickly produced a vaccine based on this bacterium, and Noguchi published a paper describing the supposedly successful vaccination of 7,964 subjects. But others found his record-keeping and statistical methods to be sloppy, and they could not replicate his results when they tried the same vaccination. In 1926, the eventual Nobel Prize–winner Max Theiler helped to prove Noguchi wrong, and the Rockefeller Institute quietly stopped making the vaccine. Noguchi tried again to prove his bacterial theory by joining a new yellow fever commission in Africa. Although he was unsuccessful in finding his bacteria, that commission did make observations that would provide a breakthrough in research. They discovered that though African monkeys carried the virus when injected with blood from human yellow fever victims, the animals did not fall prey to the disease. The African monkeys were acting as a *reservoir*: the virus in their bloodstream could get picked up when they were bitten by mosquitoes, and the mosquitoes could then transmit it to humans. However, this was not true of all monkeys—rhesus monkeys from Asia did get ill when inoculated with human yellow fever blood. At last, a nonhuman way to study the infection had been found, but this research was cut short when Noguchi and two other international researchers died of yellow fever while in Africa.

———

7   H. Noguchi, "Etiology of Yellow Fever: VII. Demonstration of Leptospira Icteroides in the Blood, Tissues, and Urine of Yellow Fever Patients and of Animals Experimentally Infected with the Organism," *Journal of Experimental Medicine* 30, no. 2 (1919): 87–93.

The next phase of investigation was taken up by Max Theiler. Max Theiler's career made him an unlikely prospect for the Nobel Prize. Born in South Africa of Swiss parents, he studied at the Royal College of Physicians and at the London School of Tropical Medicine and Hygiene, but he never earned a medical degree or a doctorate in any field. Nevertheless he found positions at Johns Hopkins, Harvard, and then the Rockefeller Institute.

In the US, Theiler heard about rhesus monkeys from the African commission. Theiler took blood from rhesus monkeys ill with yellow fever and injected it into mouse brains. The mice did not get yellow fever, but material drawn from the infected mouse brain still caused yellow fever when passed back into rhesus monkeys. In the course of this work, Theiler himself contracted yellow fever but survived it.

Theiler kept passing the infected material through consecutive mouse brains, with each passage weakening the virus's ability to cause liver damage in the recipient monkeys. This process is called *attenuation*. With increasing attenuation the virus grew weaker, and eventually it did not cause yellow fever in monkeys; instead they developed immunity.

The discovery of yellow fever in monkeys settled another piece of the puzzle. In Africa, mosquitoes transmit the virus to humans and monkeys. While bitten humans might get ill, African monkeys only carry the virus. Mosquitoes bite both humans and African monkeys, and this serves to swap the virus around: human to mosquito to human; human to mosquito to monkey; monkey to mosquito to monkey; monkey to mosquito to human.

Theiler perfected the vaccine after one hundred passages through different mouse brains for a very attenuated product—a virus weak enough to not cause yellow fever, but just strong enough to provoke an immune response in most subjects.[8] Theiler created a vaccine that still contained live virus but had little propensity to cause liver damage and yellow jaundice.

---

8   M. Theiler and H. H. Smith, "The Effect of Prolonged Cultivation In Vitro upon the Pathogenicity of Yellow Fever Virus," *Journal of Experimental Medicine* 65, no. 6 (1937): 767–86.

Unfortunately, it could cause brain damage in humans, having been grown in the brain tissue of mice. It also could not survive well outside the body unless human serum was added to the vaccine. Experimental subjects who were vaccinated sometimes got sick from the human serum in the vaccine, and some had severe neurological complications from the virus having been grown in mouse brains.[9]

Researchers at the Pasteur Institute of Tunis made a different yellow fever vaccine by omitting the human serum, hoping this would have fewer side effects. They also coated the vaccine with egg yolk, with the idea that this would cause a slow release of the vaccine and be safer yet. This became known as the French vaccine. At that time, the physician-missionary Dr. Albert Schweitzer was working at his charity hospital in Gabon, where he saw yellow fever victims regularly. When the French vaccine was still experimental, Schweitzer offered himself as a test subject. He survived that vaccination with no ill effects and then started giving it to his patients. Schweitzer became the 1952 winner of the Nobel Peace Prize for his philosophy of reverence for life, exemplified by his humanitarian work in Africa. By today's standards, his vaccination use would not be allowed, due to a lack of sufficient safety studies and the complete absence of any opportunity for the subjects to give meaningful informed consent. By the end of 1945, largely through the efforts of the Pasteur Institute, sixteen million Africans were inoculated with the French version of the vaccine. Neurological side effects were reported to be rare; however, there was no attempt to formally document side effects or inquire about long-term consequences. Compared to other versions, the French vaccine was ultimately shown to have a significantly higher rate of serious neurological effects, but it took until 1982 for it to be abandoned in Africa. No one was held accountable.

In further attempts to make an American version of the vaccine less toxic to the nervous system, Theiler's team stumbled upon a mutant—one of the viruses in their lab underwent a chance mutation resulting in a version that just happened to be less toxic to the brain.

---

9   J. E. Staples and T. P. Monath, "Yellow Fever: 100 Years of Discovery," *JAMA* 300, no. 8 (2008): 960–62.

This became the basis for a new vaccine, and it is for this vaccine that Theiler won the 1951 Nobel Prize.

———————

By 1942 some seven million doses of the vaccine produced at the Rockefeller Foundation had been given worldwide, including the inoculation of all US military recruits for WWII (the latter provided for free by the foundation).[10] The troops were generally protected from yellow fever, but within two to five months of the vaccine being administered, there were 26,771 cases of jaundice in military personnel, and many cases resulted in death.[11] Some estimates were that twenty times more soldiers had fallen victim to the vaccine than had died so far in battle. This was eventually discovered to be due to the serum used to stabilize the vaccine having been collected from donors who had hepatitis B. A follow-up study published in 1987 estimated that 330,000 men were infected with or exposed to the hepatitis B virus over a six-month period in 1941 and 1942, and about one in seven of them became sick.[12] This calamity immediately resulted in the production of an American version of serum-free vaccine. Another scare came in 1962 when a lot of vaccine that had been widely distributed in England and the US was found to be infected with a bird virus that causes cancer (avian leukosis virus). The British addressed this by adding a virus antibody to the vaccine in order to neutralize the contaminant virus. In the US it was addressed by ultrafiltering the vaccine to hopefully remove the bad virus. Then a study was done on 2,659 military vaccine recipients who'd later gotten cancer, and the bird virus was not detected in any of

10 Rockefeller Foundation, "Rockefeller Foundation Provides Yellow Fever Vaccine Free to Government for American Armed Forces" (press release), March 19, 1942, resource.rockarch.org/story/the-long-road-to-the-yellow-fever-vaccine/.

11 G. Freeman, "Epidemiology and Incubation Period of Jaundice Following Yellow Fever Vaccination," *American Journal of Tropical Medicine and Hygiene* 26 (1946): 15–32.

12 L. B. Seeff et al., "A Serologic Follow-up of the 1942 Epidemic of Post-Vaccination Hepatitis in the United States Army," *New England Journal of Medicine* 316, no. 16 (1987): 965–70.

the tumors. It was concluded that even if the vaccine were carrying the bird virus, it had not resulted in human harm.[13]

Avian Leucosis Virus in Vaccines

Most vaccines are grown on chicken embryos, and other vaccines have since been found to be contaminated with avian leukosis virus, including measles vaccine and mumps vaccine.[14]

The problem has not gone away. A 2003 study by the US Centers for Disease Control detected avian viruses in yellow fever vaccines manufactured by three different pharmaceutical companies. A follow-up of forty-three vaccine recipients satisfied the researchers that none of the humans were infected with the potentially cancer-causing bird virus. Many question if this study of forty-three subjects provides sufficient reassurance for a vaccine that is given an estimated eighty million times per year.[15] One of the current yellow fever vaccines in wide use is derived from the US version called 17DD, and it is serum free, eliminating the concern for carrying other human viruses, like hepatitis B or HIV. The 17DD vaccine-strain yellow fever virus is propagated in egg yolks that are certified to be free of avian virus. The world's largest manufacturer is the pharmaceutical giant Sanofi Pasteur. Serious neurological side effects rarely occur, but only ten cases are officially admitted by the manufacturer. The actual number is probably ten- to one-hundred-fold higher, following the usual statistics on known underreporting. A 2014 study in Brazil showed the rate of neurologic disease associated with yellow fever vaccine was 2 to 3 per 100,000

---

13 F. Piraino et al., "Serologic Survey of Man for Avian Leukosis Virus Infection," *Journal of Immunology* 98, no. 4 (1967): 702–6.

14 J. A. Johnson and W. Heneine, "Characterization of Endogenous Avian Leukosis Viruses in Chicken Embryonic Fibroblast Substrates Used in Production of Measles and Mumps Vaccines," *Journal of Virology* 75, no. 8 (April 2001): 3605–12.

15 A. I. Hussain Al, J. A. Johnson, M. Da Silva Freire, W. Heneine, "Identification and Characterization of Avian Retroviruses in Chicken Embryo-Derived Yellow Fever Vaccines: Investigation of Transmission to Vaccine Recipients," *Journal of Virology* 77, no. 2 (2003): 1105–11.

vaccine doses in the state of Rio Grande do Sul, and 0.83 per 100,000 doses in the nation.[16]

A gruesome firsthand account of the horrifying neurological effects of modern yellow fever vaccine is found in the documentary *Malcolm Is a Little Unwell*, which chronicles the postvaccination descent into madness of Malcolm Brabant, an award-winning BBC journalist. Over the course of two years following yellow fever vaccination, Brabant was institutionalized three times, sedated with powerful drugs, and subjected to electroshock before crawling to recovery. The documentary relates that Sanofi Pasteur unofficially admitted to four hundred cases of serious neurological effects. In 2019, a leading cancer expert in the UK died suddenly of total organ failure following a yellow fever vaccination.[17] There are estimated to be just under 4 serious adverse effects from the vaccine per 100,000 doses, but the risk of serious adverse effect goes up to 6.5 per 100,000 for those aged between 60 and 69, and 10.3 per 100,000 for those aged 70 and above. These numbers remain far lower than the risk of death from yellow fever.

Yellow fever vaccination is currently recommended for travelers to tropical and subtropical Africa, from Senegal on the west coast across to Ethiopia in the east, and as far south as Angola, with many countries requiring proof of vaccination with a yellow fever certificate at the border. Vaccination is also recommended for travelers to South American countries from Columbia across the continent and down, including all of Brazil.[18] In 2014, the World Health Organization announced that a single dose of yellow fever vaccine is sufficient to confer sustained immunity and lifelong protection against the disease, and that a booster dose is not needed.

---

16 R. de Menezes Martins et al., "Adverse Events Following Yellow Fever Immunization: Report and Analysis of 67 Neurological Cases in Brazil," *Vaccine* 32, no. 49 (2014) 6676–82.

17 "Royal Marsden's Leading Cancer Expert Martin Gore Dies," BBC News online, January 11, 2019, bbc.com/news/health-46836877.

18 "Yellow Fever Vaccine Recommendations," Centers for Disease Control, cdc.gov/yellowfever/vaccine/vaccine-recommendations.html.

2

Dirty Business

The 1952 Nobel Prize in Medicine went to Selman Waksman for the discovery of streptomycin, the first antibiotic effective against tuberculosis. However, streptomycin was actually discovered by Albert Schatz, a graduate student working largely independently in Waksman's laboratory. How did this happen?

Selman Waksman was born in the Ukraine region in czarist Russia and immigrated to America in 1910, obtaining his bachelor's and master's degrees in agriculture at Rutgers University in New Jersey. He discovered the role of microbes in degrading plant matter in the formation of soil, and he clarified the influences of nitrogen, minerals, and temperature on the quality of soil.[1] Waksman's careful work to elucidate the factors that make dirt good for growing contributed hugely to the next hundred years of soil-conservation efforts around the world.

Waksman had developed a systematic six-step technique for isolating pure strains of microbes from soil. He studied how some bacteria inhibit the growth of other microbes in nature's ongoing "survival of the strongest" contest for microscopic territory in dirt. Solving the composition of dirt for all mankind is a pretty big deal, but—let's face

1   S. A. Waksman and J. P. Martin, "The Role of Microorganisms in the Conservation of the Soil" *Science* 90, no. 2335 (1939): 304–5.

it—soil is not a sexy topic, and Waksman's early research did not attract the attention of the Nobel Committee.

The Nobel Prize rules for medicine state that it is to be awarded "to those who, during the preceding year, shall have conferred the greatest benefit on mankind." What could create the greatest benefit for mankind in a one-year period? There had already been four previous prizes given for the treatment of infectious disease, including the recently celebrated introductions of sulfa antibiotics and penicillin.[2]

Waksman and his research team made a study of certain soil bacteria that killed off competing species by producing toxic substances— these are called antibiotics. Although the team identified four likely antibiotics, each of them was found to be too toxic on lab animals to be considered for testing in humans. The search continued for an antibiotic from a soil organism that could be effective against bacteria but not kill the patient.

Although penicillin had licked scarlet fever, strep throat, and many other illnesses, it was not effective at combating several common conditions. Specifically, penicillin could not treat infections caused by rod-shaped bacteria that were differentiated under the microscope by staying pale when stained with Gram's method. This earned them the name gram-negative bacteria. Various species of gram-negative rods were causing tuberculosis, bladder and kidney infections, bowel infections, pneumonia, gonorrhea, and meningitis, and they were major causes of deadly surgical infections. Sulfa antibiotics could treat some gram-negative bacterial infections but not tuberculosis.

Tuberculosis (TB) is a highly infectious bacterial disease that at one time was a leading cause of death in the Western world. It is easily spread by the cough of an infected person. The incidence of TB in Western countries began declining in the nineteenth century, well

---

2   The 1945 Nobel Prize in Medicine was given to Alexander Fleming, Howard Walter Florey, and Ernst Boris Chain for the discovery of penicillin, which was actually a rediscovery of an antibiotic as documented in 1897 by Ernest Duchesne, a medical student working at the Pasteur Institute. Duchesne died of tuberculosis at the age of thirty-seven, in obscurity.

before the antibiotic era, and it also gradually became less deadly, for reasons that are not known. In the first half of the twentieth century, the two world wars caused slight bumps in TB incidence, but the overall downward trend continued; later in the century, AIDS would also cause a transient spike in TB cases. In Waksman's era, some notable persons were still dying of TB, including American author Thomas Wolfe, who died from TB of the brain in 1938. Ho Chi Minh had TB in the 1930s but recovered to become the first president of the Democratic Republic of Vietnam in 1945. Academy Award–winning actress Vivian Leigh and classical musician Igor Stravinsky also survived TB.

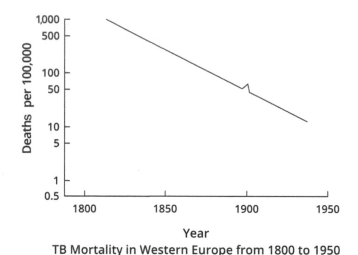

TB Mortality in Western Europe from 1800 to 1950

Albert Schatz was a graduate student working under Waksman's academic supervision in his labs. Schatz was born in Connecticut and spent his early years on the family farm. He went to college at Rutgers to study agriculture and graduated in 1942 with a major in pedology, that is, soil science (from the Greek *pedon*—ground). He was accepted as a graduate student into Waksman's laboratory, but in November that year Schatz was drafted into the US Army. He was stationed as a hospital microbiologist at military facilities in Florida that received the wounded returning from the front lines. There, Schatz saw many men his own age dying from untreatable gram-negative infections. He gained a great deal of experience in handling infected blood samples and culturing them in the lab, where he carefully tested them for susceptibility to the few antibiotics that were available. Schatz was released from the service in May of 1943, and he returned to the Rutgers lab.

Meanwhile, two Mayo Clinic doctors, William Feldman and H. Corwin Hinshaw, had asked Waksman to collaborate on antibiotic studies against the bacterium that causes TB, tubercle bacillus. While Waksman had the microbiology lab, Feldman and Hinshaw had the facilities to do testing in animals and humans.[3] Schatz readily volunteered to work on the TB project, and he secured a culture of tubercle bacillus from Feldman that was reported to be responsible for the most lethal TB in America.

In a firsthand account, Schatz relates that he was relegated to the basement lab for this work: "Waksman assigned me to work in the basement laboratory, because he wanted to be as far away from the tubercle bacillus as he could be. He was deathly afraid of catching the disease. That is why he never once visited me in the basement laboratory during the entire time of my research."[4] Infection was a well-founded concern for researchers of Waksman's generation, as it had not been uncommon for scientists to fall prey to the substances they studied, including six early Nobel laureates in medicine who suffered from TB.[5] However, Schatz was of a younger generation, and he had gained extensive experience in safe handling of infectious material during his army hospital stint.

---

3   H. C. Hinshaw, M. M. Pyle, and W. H. Feldman, "Streptomycin in Tuberculosis," *American Journal of Medicine* 2, no. 5: 429–35.

4   A. Schatz and I. Auerbacher, *Finding Dr. Schatz: The Discovery of Streptomycin and a Life it Saved* (2006).

5   See M. Dolan, *Boneheads & Brainiacs: Heroes and Scoundrels of the Nobel Prize in Medicine* (Fresno, CA: Quill Driver Books, 2020) (hereafter cited as *Boneheads & Brainiacs*), 217.

Waksman kept his distance two floors up and although there was no direct supervision, Schatz regularly informed his academic boss of the progress of his work. Unbeknownst to Schatz, Waksman had entered into a paid consulting agreement with the pharmaceutical company Merck, who wanted to be kept abreast of any new discoveries they might commercially exploit.

Schatz found two strains of bacteria that made an antibiotic substance that was given the name "streptomycin." The source bacterium was subsequently named *Streptomyces griseus*. One of the *S. griseus* strains was from soil, for which Schatz used his own isolation technique rather than Waksman's six-step procedure. The other *S. griseus* strain was from a pure culture that a colleague from the Poultry Science Department had given him, derived from the throat of a healthy chicken. Both produced streptomycin that killed the deadly tuberculosis bacteria obtained from the Mayo Clinic, but Schatz focused his attention on the stronger antibiotic from the chicken-sourced strain. Interestingly, Waksman had isolated *S. griseus* from soil in 1916, but the particular strain he'd found did not make any antibiotic.

Schatz grew the chicken-sourced *S. griseus* and isolated the streptomycin it produced, and Waksman had it sent to Feldman at the Mayo Clinic. There, animal tests did not show any severe toxicity, and subsequent human tests on patients with tuberculosis were successful. In 1944, Schatz published two papers on the discovery in which he was listed as the senior author.[67] Lead authorship on scientific papers is not just a symbol of prestige; it implies that the first-listed author is the major contributor to the work. Waksman was listed as the last of three authors on the first paper, and the second of two authors on the second

---

6   A. Schatz, E. Bugie, and S. A. Waksman, "Streptomycin, a Substance Exhibiting Antibiotic Activity against Gram Positive and Gram Negative Bacteria," *Proceedings of the Society for Experimental Biology and Medicine* 55, no. 1 (1944): 66–69.

7   A. Schatz and S. A. Waksman, "Effect of Streptomycin and Other Antibiotic Substances upon Mycobacterium Tuberculosis and Related Organisms," *Proceedings of the Society for Experimental Biology and Medicine* 57, no. 2 (1944): 244–48.

paper. It was also unusual for academic supervisors to get author credit when they had not stepped foot in the research lab.

In 1945, Schatz wrote his PhD thesis on the discovery of streptomycin,[8] and also presented his discovery at the proceedings of the National Academy of Science. On January 31, 1945, Waksman brought patent-application papers to Schatz. These included an oath and an affidavit, each signed by Waksman, that clearly state Schatz and Waksman jointly discovered the antibiotic. On May 6, 1946, Waksman got Schatz to sign away his patent rights, assigning them to the Rutgers Research and Endowment Foundation for a token payment of one dollar. Schatz presumed they were jointly giving up rights to the foundation for the good of mankind. Schatz had no way of knowing that Waksman had entered into a separate agreement with the foundation to retain 20 percent of the royalties, nor would he have reason to suspect that Waksman was making arrangements for his personal gain.[9]

Schatz graduated with his PhD in 1945 and moved on from Rutgers. The streptomycin patent was granted in 1948. Waksman sent more papers to Schatz to sign, this time assigning royalties from foreign sales of streptomycin to the foundation. Waksman and the foundation assured Schatz that no individual was earning royalties. Schatz did not have any suspicions until Waksman was featured on the cover of *Time* magazine on November 7, 1949, wherein he was lauded as the sole discoverer of streptomycin. Schatz soon discovered that Waksman had been receiving royalties the whole time.

Schatz tried to work it out with Waksman in a gentlemanly way, but Waksman disregarded the requests for negotiations. Schatz brought suit in order get due credit as the discoverer, a share of the royalties, and restitution for fraud and deception by both Waksman and Rutgers Research and Endowment Foundation. In today's lawsuit-happy

8   A. Schatz, "Streptomycin, an Antibiotic Agent Produced by Actinomyces griseus," PhD thesis, Rutgers University, July 1945.

9   A. Schatz and I. Auerbacher, *Finding Dr. Schatz: The Discovery of Streptomycin and a Life it Saved* (2006).

society this may not seem outrageous, but in the hallowed halls of academia in 1950, it was an extremely aggressive move.

Pretrial depositions revealed that Waksman had received $350,000 in royalties, which he had previously publicly denied, while the foundation had gained $2.6 million from patent rights on Schatz's discovery of streptomycin. The case dragged on at great cost to Schatz and was finally settled out of court in December 1950. Schatz was granted credit, legally and scientifically, as the codiscoverer of streptomycin. He was to receive 3 percent of royalties, while Waksman was allowed to receive 10 percent, with additional royalties also going to a number of other former laboratory students and associates, including the lab's glassware dishwasher. This latter part of the agreement was an attempt to discount Schatz's contribution by equating it to the work of anyone else present in the lab. The foundation was to use part of the funds they earned from streptomycin to establish an institute of microbiology (later named in honor of Waksman). As part of the agreement, Schatz was required to drop the charges of fraud and deception.

It all seemed settled, but Schatz would continue to be the loser on two more fronts. First, he had a very hard time securing an academic position, having been effectively blacklisted as a litigation-happy liability. The next blow came in October 1952, when the Karolinska Institute announced Selman Waksman as the sole Nobel Prize winner in medicine "for his discovery of streptomycin." Schatz wrote in protest to the Nobel organization, supported by additional letters from a very few of his academic colleagues. The "old boys club" was not willing to rock that boat, most likely because many had similar situations of failure to adequately credit their subordinates—those who did the actual laboratory work.

Despite the legal decision that fully recognized Schatz as equal codiscoverer of streptomycin occurring nearly two years before the Nobel announcement, the Nobel organization adhered to their policy to never go back on a selection. Their only concession was to soft-pedal the affair by the time of the December awards ceremony.

Karolinska committee member A. Wallgren announced the prize to Waksman, "For your ingenious, systematic and successful studies of the soil microbes that *have led to the* discovery of streptomycin" (italics added).[10] In his Nobel lecture, Waksman gave Schatz no credit, and only mentioned "A. Schatz" in the middle of two dozen other names at the end of his speech.[11] Waksman subsequently enjoyed a long career at Rutgers and died in 1973 at the age of eighty-five.

Albert Schatz authored three books and over five hundred scientific papers, and he finished his career at Temple University. In April 1994, on the fiftieth anniversary of the discovery of streptomycin, Schatz was awarded the Rutgers Medal, the university's highest honor, for his codiscovery of streptomycin. He died in 2005 at the age of eighty-five.

In 1950, future rock stars Tom Jones and Ringo Starr were each twelve years old when they both came down with tuberculosis. They were presumably treated with a protracted series of painful strepto-mycin injections. Jones may have suffered one of the early instances of streptomycin-resistant TB, because he spent two years in a hospital. Also in 1950, English author George Orwell died of a ruptured lung artery from TB, another case of probable drug resistance. In most parts of the world the tubercle bacillus has successfully mutated to develop resistance to streptomycin. It is no longer used as first-line treatment. Today tuberculosis is routinely treated with a four-drug regimen of isoniazid, rifampin, ethambutol, and pyrazinamide.[12]

10 As quoted in A. Schatz and I. Auerbacher, *Finding Dr. Schatz: The Discovery of Streptomycin and a Life it Saved* (2006).

11 S. Waksman, "Streptomycin: Background, Isolation, Properties, and Utilization," December 12, 1952, NobelPrize.org, nobelprize.org/prizes/medicine/1952/waksman/lecture/.

12 "Global Tuberculosis Report 2019," World Health Organization, 2019, who.int/publications/i/item/9789241565714.

3

Squiggles and Gold

Hans Krebs and Fritz Lipmann shared the 1953 Nobel Prize in Medicine for their separate but related discoveries about basic metabolism.

Hans Adolf Krebs was born in 1900 in Germany. He made his first important discovery by sorting out exactly how the body neutralizes the toxic ammonia that is left over from the metabolism of protein.[1] In 1932 the dean of the medical faculty at the University of Freiburg, surgeon E. Rehn, praised Krebs as an outstanding scientist whom the university should regard with pride. Krebs's work was abruptly interrupted in 1933 when Germany's National Socialist government passed the Law for the Restoration of the Professional Civil Service, which scrubbed all public institutions of Jews and anti-Nazis.[2] On April 12, 1933, the same Dean Rehn called for Krebs's immediate suspension. A letter from the minister of education explained the termination: "The

---

1 H. Krebs and K. "Henseleit, Untersuchungen über die Harnstoffbildung im Tierkörper" [Studies on urea formation in the animal body], Hoppe-Seyler's Zeitschift für physiologische 210 (1932): 33–66.

2 The legal document describing Law for the Restoration of the Professional Civil Service can be seen on the website of the United States Memorial Holocaust Museum, section "Timeline of Events 1933–1938," at ushmm.org/learn/timeline-of-events/1933-1938/law-for-the-restoration-of-the-professional-civil-service.

Minister of the Interior has decided that all members of the Jewish race (regardless of their religion) who are employed in the civil service or educational facilities will be suspended until a further decision is made. All academic instructors and assistants in this category are to be informed that they are immediately suspended."[3]

Krebs was fired from his academic post and became a scientific refugee. The Academic Assistance Council (AAC) came to his aid. The AAC was founded in May 1933 by William Beveridge, then the director of the London School of Economics. Beveridge formed the organization as a response to the Nazi decree, with initial support and funding coming from British academics. With help from the AAC, Krebs was invited to Cambridge.[4] Krebs's laboratory equipment had been seized when he was fired, but in June the entire contents of his old lab were acquired with funds from the Emergency Association of the German Sciences and the Rockefeller Foundation. His loyal former coworkers packed his instruments for export to Cambridge. On his travel papers, Krebs indicated his religion as "non-denominational," and he officially carried only the allowed ten marks in cash, but hid another five hundred marks within his books.

By 1936, it became clear that the need for ongoing assistance was not temporary, and the AAC joined the Swiss-based Emergency Association of German Scientists Abroad to become a new larger organization called the Society for the Protection of Science and Learning (SPSL).[5] Its stated goal was "the brotherhood of scientific endeavour, regardless of race and creed and politics: and it stands for it, not by passing pious resolutions or by putting out disguised political propaganda,

3   Roth, Klaus, "Sir Hans Adolf Krebs (1900–1981), Part 2." *Chemie in unserer Zeit* [*Chemistry In Our Time*] (Wiley-VCH, December 1, 2020. DOI: 10.1002/chemv.202000121).

4   For more detailed information on the role of Cambridge as an academic refuge, see, "Keynote: Sir Hans Krebs and the Sir Hans Krebs Trust. Introduced By Stephen Wordsworth," Trinity College, Cambridge, May 7, 2020, youtube.com/watch?v=QYoFCHD8_pc&feature=emb_logo.

5   E. Rutherford, "The Society for the Protection of Science and Learning," *Science*, April 17, 1936, 372.

but by trying to help colleagues in their need."[6] In the early years, some two thousand people were saved, sixteen of whom would become Nobel Prize winners.

At Cambridge, Krebs studied how cells convert fuel to energy. He discovered that after food gets broken down into sugars, fats, and proteins, these components use oxygen in a multistep chain reaction. The process converts the biochemicals into energy-providing molecules (ATP), with leftover carbon dioxide and water. This became known as the Krebs cycle. The ATP supplies the power for all the work of the body, such as muscle movement, breathing, and cell function.

In 1937, Krebs submitted a report on his discovery to the editors of *Nature*, the world's premier scientific journal, but it was rejected.[7] He managed to get his work published in a small Dutch biochemistry journal, and it was soon recognized the world over as groundbreaking.[8]

Krebs soon became a naturalized British citizen, and he served on the faculty of the University of Sheffield for many years. In 1970, there was much debate about cutbacks to the US federal budget for scientific research. Dr. Krebs was reported to have told a meeting of the American Philosophical Society that, among other things, the government should eliminate wasteful and unproductive research, which he characterized as "occupational therapy for the university staff."[9] The last part of his career was at Oxford, where Krebs remained highly active with the publication of hundreds of scientific articles and contributions to textbooks. Hans Krebs died in 1981 at the age of eighty-one.

----

6   "Science and Learning in Distress," *Nature* (1938): 1051–52.

7   F. MacDonald, "8 Scientific Papers That Were Rejected before Going on to Win a Nobel Prize," *Science Alert*, August 19, 2016, sciencealert.com/these-8-papers-were-rejected-before-going-on-to-win-the-nobel-prize.

8   H. A. Krebs and W. A. Johnson, "The Role of Citric Acid in Intermediate Metabolism in Animal Tissues," *Enzymologia* 4 (1937): 148.

9   As quoted in "Sir Hans Krebs, Winner of Nobel for Research on Food Cycles, Dies," *New York Times*, December 9, 1981.

The work of the organization that had assisted Krebs to flee to England did not end after the second world war. The SPSL assisted academic refuges from Stalin's regimes in the USSR and Eastern Europe, and in later years it assisted academics threatened by military juntas in Chile and Argentina or apartheid in South Africa and those who were persecuted for religious, racist, or political reasons in other African countries as well as in Syria and Iraq. In 1999, the SPSL changed its name to the Council for Assisting Refugee Academics (CARA). This name was modified again in 2014 to become the Council for At-Risk Academics.

In 2015, Krebs's descendants arranged for Sotheby's to auction his Nobel medal. It was made of 23-karat gold, measured 6.5 centimeters in diameter, weighed 196 grams, and was inscribed, "Inventas vitam juvat excoluisse per artes" (the word-for-word translation is "inventions enhance life which is beautified through art"). Krebs's medal was sold to an anonymous collector for $275,000.[10] The money went toward the Sir Hans Krebs Trust, which supported young biomedical scientists working in the UK who have been forced to leave their own country because of conflict, discrimination, or danger. The trust subsequently partnered with the Council for At-Risk Academics to administer scholarships for refugees.[11]

When it was revealed that the buyer of the medal was planning to take his purchase out of the country, England's Reviewing Committee on the Export of Works of Art and Objects of Cultural Interest (RCEWA) managed to get a temporary stay on the movement of the medal. RCEWA felt the object should remain in the UK as a testament to the contribution of scientific refugees to the war effort. They got a two-month hold while an English buyer was sought. In 2016, the medal went on the auction block at Nate D. Sanders, where it sold for $269,000.

---

10 A. B. Kenner, "Krebs Nobel Auctioned," *Scientist*, July 16, 2015.

11 "Who We Are," Council for at Risk Academics, cara.ngo/who-we-are/our-history/.

Nobel Gold

Krebs's medal was 23-karat gold, but since 1980 the medals have been made of an 18-karat green-gold core, which is an alloy of gold and silver with trace amounts of copper, then plated with 24-karat gold. Nobel Prize medals have fetched a variety of amounts on the auction block. In 2008 the Nobel Peace Prize medal won by Aristide Briand in 1926 was sold for $14,000. The medal awarded in 1962 to Francis Crick for his DNA discoveries sold in 2013 for $2.27 million. The proceeds went to the Francis Crick Institute for medical research in London. The medal of his cowinner, James Watson, sold the following year for $4.1 million. The Russian billionaire who bought it, Alisher Usmanov, gave it back to Watson. In 2015, Leon Lederman, physics winner from 1988, sold his medal to pay mounting medical bills, fetching $765,000. He died the following year. That same year, the medal for the 1963 medicine prize awarded to Alan Lloyd Hodgkin sold for $795,614. Also in 2015, the medal for the 1971 economics prize (not strictly a Nobel Prize) awarded to Simon Kuznets brought in $390,848. The medal for the 1994 Nobel Prize in economics awarded to John Nash sold for $735,000 in 2016, a year after he'd died in a car accident. The medal of the 1965 physics prizewinner Richard Feynman sold for almost $1 million in 2018.

The cowinner in 1953 was Fritz Albert Lipmann, who was born in 1899 in Königsberg, then the capital of East Prussia. The region is now called Kaliningrad, and is an outpost of the Russian Federation sandwiched between Poland and Lithuania. Even before getting his medical degree in Berlin, Lipmann was conducting basic research in chemistry. He continued with biochemistry research to earn his PhD at the Kaiser Wilhelm Institute for Biology, and it was there he met Krebs in 1927.

Lipmann subsequently left Germany to work in research labs in New York, Copenhagen, and London.[12]

Lipmann sought to answer exactly how the metabolic products of glucose were pulled into the Krebs cycle—there was a missing intermediate chemical. He discovered the small linking molecule and named it coenzyme A. The role of a coenzyme is to hold atoms or molecules together to make it possible for enzymes to work. Two years later, biochemist Feodor Lynen detected yet another intermediate substance, but his contribution was not recognized with a Nobel Prize until 1964. These combined discoveries revealed how the body converts chemical energy into physical energy. They are the underpinnings for all subsequent work on metabolism and nutrition, and they also made it possible for others to discover how DNA is made.

Lipmann immigrated to America and worked at the Marine Biological Laboratory at Woods Hole on Cape Cod in Massachusetts, Cornell Medical School in New York, and Massachusetts General Hospital in Boston. He did not meet Krebs again until they were receiving their shared prize at the ceremonies in Sweden in 1953.

Lipmann subsequently contributed to many biochemical discoveries, but his other claim to fame was the introduction of the squiggle "~" to biochemistry notations, known on the computer keyboard as the tilde. He used the squiggle to denote that a molecule had a high-energy bond. Lipmann used this shorthand for the ATP molecule, AMP~P~P, with the squiggles indicating that when those bonds are broken, they release a lot of energy.[13]

Chemists pounced on Lipmann's squiggle as inaccurate and misleading, but it was such a handy way to describe the concept that it was used in textbooks for decades to come. The squiggle is not used anymore, because it is now known that it is neither the ATP molecule

12 W. P. Jenck and R. V. Wolfenden, *Fritz Albert Lipmann, 1899–1986: A Biographical Memoir* (Washington, DC: National Academy Of Sciences, 2006).

13 F. Lipmann, "Metabolic Generation and Utilization of Phosphate Bond Energy," *Advances in Enzymology and Related Areas of Molecular Biology* 1 (1941): 99–162.

nor the bonds in the molecule that are high energy. Rather, it is the microenvironment (the system) in which the ATP is operating that makes it "molecular energy currency" for the cells. Most ordinary people understand the idea inherent in the squiggle better than the concept of energy currency, but then again, the job of the scientist is not to make things clear to ordinary folks.[14]

In 1986, at the age of eighty-seven, Lipmann was notified that his latest research grant application had been successful. He was preparing notes for a scientific conference presentation when he uttered, "I can't function anymore," and slipped into a coma, dying shortly thereafter.[15]

14 H. Kleinkauf, H. Dohren, and L. Jaenicke, *The Roots of Modern Biochemistry: Fritz Lippmann's Squiggle and Its Consequences* (Berlin: De Gruyter, 2011).

15 As quoted in H. Kleinkauf, H. Dohren, and L. Jaenicke, *The Roots of Modern Biochemistry: Fritz Lippmann's Squiggle and Its Consequences* (Berlin: De Gruyter, 2011).

4

The Polio Researchers

The 1954 Nobel Prize was shared by John Enders, Thomas Weller, and Frederick Robbins for their research on poliovirus.

John Franklin Enders was the son of John Ostrom Enders, the CEO of Hartford National Bank. The young Enders's wandering educational path did not seem likely to lead to a distinguished career, much less a Nobel Prize. He interrupted his studies at Yale for a three-year stint with the Army Air Corps Reserves. He returned to complete his degree at Yale and then dabbled in real estate. Not surprisingly, this soon-to-be-brilliant researcher found real estate too dull. Enders enrolled at Harvard with the thought of becoming a teacher, but after earning his master of arts degree, he did not immediately continue his studies.

Enders eventually befriended a roommate who was an instructor in microbiology. "I fell into the habit of going to the laboratory with him in the evening and watching him work," Enders recalled. "I became increasingly fascinated by the subject—which manifestly gave him so much pleasure and about which he talked with such enthusiasm—and so eventually decided to change the direction of my studies."[1] Enders

1   As quoted in E. Ofgang, "How a CT Man who Majored in English at Yale Became the 'Father of Modern Vaccines,'" *Connecticut Magazine*, Sepember 8, 2020.

was soon accepted into Harvard's microbiology doctoral program under the supervision of the famous microbiologist Hans Zinsser.

Hans Zinsser

Hans Zinsser was one of the most productive microbiology researchers of the century, but he was mysteriously overlooked for a Nobel Prize, although he received three nominations in 1940, the year of his death. Zinsser was the discoverer of the bacterium that caused a form of typhus carried by lice, and he developed a protective vaccine against it. He was also a prolific author, including a "biography" of typhus entitled *Rats, Lice and History* in which he wrote: "Swords, lances, arrows, machine guns, and even high explosives have had far less power over the fates of nations than the typhus louse, the plague flea, and the yellow-fever mosquito."[2] Zinsser wrote extensively on the subject of immunity in an era where little was known about the complex biochemical and cellular basics of the immune system. He made major contributions in the US Army in WWI, including risking his life to care for wounded soldiers on the front lines. However, his most wide-reaching impact in the war was keeping the troops healthy with field sanitation and hygiene practices that kept typhus and other diseases to a minimum. Zinsser continues to be well known in medicine for his *Textbook on Microbiology*. He was also an accomplished poet.[3]

The lucky beneficiary of Zinsser's enthusiasm for his work and teaching, Enders initially focused his research on bacteria. His graduate work was on how the body reacts to tuberculosis, and he next

---

2   Hans Zinsser, *Rats, Lice, and History: Being a Study in Biography, which, after Twelve Preliminary Chapters Indispensable for the Preparation of the Lay Reader, Deals with the Life History of Typhus Fever*, rev. ed. (New Brunswick, NJ: Transaction Publishers, 2007).

3   H. Zinsser, *Spring, Summer & Autumn: Poems* (New York: A.A. Knopf, 1942).

studied the *Pneumococcus* bacteria that causes common pneumonia. Enders helped to develop a method to grow *Rickettsia* bacteria, a germ that's carried by body lice and causes typhus. The growing process was refined until it produced enough quantities to manufacture a vaccine against typhus.

As an assistant professor, Enders supervised the work of medical students on the growth of *Vaccinia* (smallpox) virus in chicken tissue. One of those students was to be his future shared prizewinner, Thomas Weller. Enders collaborated on projects that grew *Influenza A*, again with the goal of making enough to produce a vaccine.

Enders was invited to head up a new infectious-disease-research unit at Boston Children's Hospital. There he studied the virus that causes mumps, assisted by a newcomer to the lab, Frederick Robbins, who would be another of his Nobel Prize cowinners. They were able to grow the mumps virus on chicken tissue, a necessary step toward making a vaccine. Next, they focused on growing the virus that causes chicken pox (*Varicella*)

Meanwhile, they inoculated four unused culture tubes with poliovirus obtained from lab animals. To their surprise, the poliovirus cultures were successful, which marked the first time poliovirus could be studied from growth on nonnerve tissues. Polio is a viral illness that primarily infects the gut. While 70 percent of people who encounter the virus never get any symptoms, 30 percent do get sick, and in 0.5 percent of cases, the virus invades the nerves of the spinal cord to cause a paralysis of limbs and/or the respiratory muscles called "poliomyelitis." Future US president Franklin D. Roosevelt contracted polio in 1921 at age thirty-nine and was left crippled by it. At the time Enders was studying polio in 1949 and 1950, there were 33,300 polio cases in the US.

Enders turned all of the lab's resources toward the study of polio. The Enders-Weller-Robbins method was the first to show that virus recovered from human polio patients could be grown in test tubes. To the normal layperson, the actual mixture sounds like the creation of a mad scientist. As the three reported in a 1949 paper: "The cultures consisted of tissue fragments suspended in 3 cc of a mixture of balanced salt solution (3 parts) and ox serum ultrafiltrate (1 part). Tissues from human embryos of 2 ½ to 4 ½ months as well as from a premature human infant of 7 months' gestation were used. These were: the tissues of the arms and legs (without the large bones), the intestine, and the brain."[4] They added a suspension of mouse brain infected with poliovirus to the broth.

Weller kept successive generations of polio cultures growing, which weakened in virulence with each iteration. This work by Enders, Robbins, and Weller made it possible for other researchers to quickly develop a polio vaccine, which was accomplished in February 1954 by Jonas Salk. But Enders had serious reservations about the methods used by Salk. Salk's preparation involved inactivating the poliovirus by treating with formaldehyde. In theory, the inactivated virus retained enough similarity to the live virus for the body to recognize it and make antibodies against it. Enders thought formaldehyde was insufficient to kill the native virus. Nevertheless, Salk was confident enough in his preparation to bring the new vaccine home and inject his three young sons. His eldest son, Peter Salk, was nine at time. In 2020, Peter Salk, then a seventy-six-year-old physician and infectious-disease specialist, cautioned against rushing the development of a COVID vaccine in an interview: "I congratulate the impulse on the part of the federal government right now to want to speed things up as much as possible," he said. "What concerns me is knowing that in the past there have been unexpected things that have taken place with vaccines that had not been foreseen."[5]

---

4   J. F. Enders, T. H. Weller, and F. C. Robbins, "Cultivation of the Lansing Strain of Poliomyelitis Virus in Cultures of Various Human Embryonic Tissues," *Science* 109 (1949): 85–87.

5   G. Myre, "Among The 1st to Get a Polio Vaccine, Peter Salk Says Don't Rush a COVID-19 Shot," Weekend Edition Saturday, NPR, May 30, 2020, npr.org/2020/05/30/861887610/among-the-1st-to-get-a-polio-vaccine-pete-salk -says-dont-rush-a-covid-19-shot.

Within a few months, the experimental Salk polio vaccine was "tried out" on over one million school-aged children, often without the opportunity for full informed consent. It was licensed a year later in 1955. At first, it was reported that fifty-one vaccine recipients became paralyzed from polio and five people died of it.[6] It was later realized that at least 260 children developed paralytic polio from the vaccine.[7] This was traced to one of the Salk vaccine lots manufactured by Cutter Pharmaceuticals being contaminated with persistent wild poliovirus—in other words, some original live virus that had not been killed. Although Cutter was held responsible, Enders was convinced the methods developed by Salk to inactivate the virus were not good enough, rather than any manufacturing problem at the pharmaceutical company.

Salk versus Sabin Polio Vaccine

The Salk inactivated poliovirus vaccine, developed in 1954 and approved in 1955, required a shot to deliver the inactivated virus. In 1961, Salk's virology competitor, Albert Sabin, developed a version of live attenuated (weakened) polio vaccine that was dosed as a sweetened liquid taken by mouth. It was cheaper to manufacture, did not require injections, was more successful at conferring immunity, and offered longer-lasting protection. Sabin's oral vaccine replaced the Salk injection for a few decades, but there was still a fraction of a percent of cases of polio arising from the Sabin vaccine strain. After improvements in the techniques of inactivating the poliovirous in the Salk style of inactivated poliovirus, it was determined to have no risk of causing polio. Since 2000, inactivated poliovirus has been the only approved polio vaccine to be given in the US.

---

6 J. Latson, "The Vaccine Everyone Wanted," *Time* , February 23, 2015.

7 N. Nathanson and A. Langmuir, "The Cutter Incident: Poliomyelitis Following Formaldehyde-Inactivated Poliovirus Vaccination in the United States during the Spring of 1955. I. Background," *American Journal of Hygiene* 78 (1963): 16–28.

From 1952 to 1958, Enders collaborated with various researchers in his lab to study the measles virus, rubeola, which he considered to be a much bigger public health problem than polio, with seven hundred thousand measles cases in 1954.[8] He wanted to remain hands-on for the measles research all the way through to actual vaccine production to avoid the problems encountered by the Salk polio vaccine.

Enders and his associate Thomas Peebles collected samples from several ill students during a measles outbreak in Boston. They succeeded in isolating measles from a child named David Edmonston, and this specific virus became the basis for a vaccine. Once again, Enders concentrated all of the lab's resources on a single project. After successfully growing the measles virus on lab plates, Enders and his team were able to propagate successive strains that were less and less virulent. They finally developed a strain that did not cause measles but did cause antibody production, thus imparting a greater chance of vaccine recipients generating immune responses to wild-strain measles virus.

David Edmonston's Measles

David Edmonston was an eleven-year-old sick with the measles when his throat swab and blood samples yielded the virus that provided the basis for the first measles vaccine. When his own son was school age in the 1970s, Edmonston did not let him get vaccinated, because his wife was "dead set against it."[9]

They first tested the measles vaccine in monkeys, and then Enders and his coresearcher Samuel Katz injected themselves to check for side effects. Broad-scale testing was similar to that for the Salk polio

---

8   Data from "CDC/MMWR Summary of Notifiable Diseases, United States, 1993" and "CDC/MMWR Summary of Notifiable Diseases, United States, 2008," as presented in "The History of Vaccines," College of Physicians of Philadelphia, historyofvaccines.org/content/graph-us-measles-cases.

9   V. Iannelli, "David Edmonston and the Measles Vaccine," Vaxopedia, October 27, 2016, vaxopedia.org/2016/10/12/david-edmonston-and-the-measles-vaccine/.

vaccine trials, including systematic administration of the experimental measles vaccine to disadvantaged populations. This included youngsters at the Fernald State School in Massachusetts, ostensibly a residential facility that cared for so called feeble-minded children. In reality, Fernald was a dumping ground where social misfits or children considered "inconvenient" were confined and often abused. Katz describes that all parents were counseled and gave informed consent for the vaccine,[10] but inside stories of Fernald from former inmates show that it was common for parents to be impossible to locate and thus unable to provide consent.[11] The children at Fernald were used for the experimental measles vaccination on the heels of a study led by the Massachusetts Institute of Technology and involving Harvard. From 1946 to 1953, under sponsorship of the Quaker Oats company, researchers studied the uptake of calcium and iron by using radioactive materials. They coerced cooperation by getting the Fernald boys to join a "science club," promising participants extra food and trips to Red Sox baseball games, then letting them feast on radioactive oatmeal and milk. This all came to light in 1993, when federal records on postwar radiation experiments were declassified. The revelations resulted in a $1.85 million settlement in 1998.[12] So the extent of actual informed consent for the experimental measles vaccine on the Fernald boys is open to speculation.

The killed-measles-virus vaccine was reported to have over 95 percent efficacy, and routine measles vaccination started in 1963. The so-called Edmonston B vaccine strain proved to have an unacceptably high rate of fever and rash, but further attenuation eventually led to the Edmonston-Moraten strain, which became the seed strain available in the United States.

---

10 S. L. Katz, J. F. Enders, and A. Holloway, "The Development and Evaluation of an Attenuated Measles Virus Vaccine," *American Journal of Public Health and the Nation's Health* 52, suppl. 2 (1962): 5–10.

11 M. D'Antonio, *The State Boys Rebellion* (New York: Simon and Schuster, 2005).

12 L. Johannes, "MIT, Quaker Oats Settle Lawsuit Involving Radioactive Experiment," *Wall Street Journal*, January 2, 1998.

Upon more widespread use in the population, the effectiveness in the real world did not live up to the efficacy claimed in the early studies. In 1967, the US adopted the use of a live-attenuated-virus strain instead of a killed virus, which gave a higher rate of actual immunization. In 2019, the CDC announced that persons who were vaccinated against measles in the early sixties might need another measles shot, as the vaccine was "known" to be less effective then.[13]

Vaccine Efficacy versus Effectiveness

Vaccine efficacy is the percent reduction in disease incidence in a test group of vaccinated subjects compared to an unvaccinated group under optimal conditions. Such subjects usually do not have any underlying medical illnesses. However, vaccine effectiveness is the ability of a vaccine to prevent outcomes of interest in the "real world." For example, a 2020 study on the flu vaccine in the UK found that flu vaccination in persons aged sixty-five to seventy-five did not reduce hospitalization rates or reduce death rates from all causes.[14] This study showed that flu vaccine may be efficacious in controlled studies, but in real life it may not be effective.

There have been some reports of vaccinated children coming down with measles caused by the live-vaccine strain; most, but not all of them, had underlying immune system illnesses that may have made

---

13 As reported in S. Rowan Kelleher, "Born in the 1960s? The CDC Says You May Need a Measles Shot before Traveling," *Forbes*, April 20, 2019.

14 M. L. Anderson, C. Dobkin, and D. Gorry, "The Effect of Influenza Vaccination for the Elderly on Hospitalization and Mortality: An Observational Study With a Regression Discontinuity Design," *Annals of Internal Medicine*, 172, no. 7 (2020): 445–52.

them susceptible.[15] However small, the risks remains real, including vaccinated kids shedding the vaccine-strain virus in their urine for up to two weeks after the jab.[16] Vaccine manufacturer Merck includes "vaccine strain varicella [measles]" in the FDA-required package insert warnings of their ProQuad vaccine.

Vaccination versus Immunization

Vaccination is the act of administering a vaccine. The goal of the vaccination is to provoke the immune system in the recipient. When successful, the vaccine causes the body to generate antibodies against the particular infecting agent (bacteria or virus). However, just because one can measure and compare the amount of antibodies to attenuated or killed bacteria or virus, it is not assured that those antibodies will be specific enough or sufficient to protect against infection with wild-strain disease. For example, the acellular pertussis vaccine has been losing its ability to cause immunity in recent decades, with up to 85 percent of whooping cough occurring in fully vaccinated people in recent outbreaks.[17]

Enders was wealthy, his father having left him with $19 million, and he never tried to patent his work or share results with the media before it underwent peer review—that is, other labs scrutinizing his data and having the opportunity to replicate the experiments to see if

15 A. Bitnun et al., "Measles Inclusion-Body Encephalitis Caused by the Vaccine Strain of Measles Virus," *Clinical Infectious Diseases* 29, no. 4 (1999): 855–61; P. Y. Iroh Tam et al., "Measles Vaccine Strain from the Skin Rash of A Digeorge Patient Receiving Tumor Necrosis Factor Inhibitor," *Pediatric Infectious Disease Journal* 33, no. 1 (2014): 117; F. Kobune et al., "Characterization of Measles Viruses Isolated after Measles Vaccination," *Vaccine* 13 no. 4 (1995): 370–72.

16 B. L. Fisher, "The Emerging Risks of Live Virus & Virus Vectored Vaccines: Vaccine Strain Virus Infection, Shedding & Transmission," National Vaccine Information Center, 2014.

17 M. R. Weigand et al., "Genomic Survey of Bordetella pertussis Diversity, United States, 2000–2013," *Emerging Infectious Diseases* 25, no. 4 (2019): 780–83.

they got the same results. Enders consistently shared his knowledge with potential competitors and even returned unspent grant money to the National Institutes of Health at the end of the year.

John Enders died in 1985 at the age of eighty-eight.

———————

Enders's student and cowinner Thomas Huckle Weller was born in 1915 in Ann Arbor, Michigan, to a father who was a pathologist. At Harvard Medical School, Weller began working as a student in the lab with John Enders. After a break to serve in the Army Medical Corps in WWII, Weller returned to Enders's lab at Children's Hospital in Boston.

Like Enders, Weller was concerned about what happened with the development of the poliovirus vaccine after the successful cultures left their lab. They didn't think that Jonas Salk's proposed method—rinsing the cultures in formaldehyde to kill off all of the virus, leaving just enough dead material to provoke an immune response—would be effective. However, at that time, as in 2020 with the rush to make a COVID vaccine, there was incredible public demand and intense political pressure to come up with a vaccine. Meanwhile, reports of vaccinated kids getting polio started to appear. In a letter of June 29, 1956, Weller wrote to another virologist who was concerned about the safety of Salk's vaccine,

> I have been extremely upset by the moral and ethical aspects of the manner in which the whole vaccine program has been handled. One cannot deny that there have been pressures exerted at very high levels on public officials and on scientists that have molded opinions. Also, it is very clear that the deliberate selection of proponents on advisory committees has tended to affect the tone of official announcements. I cannot remember at any time being aware of a situation in which there has been such a

close approximation to an "official party line" in a matter of scientific import.[18]

Weller's concerns did not stop the Salk vaccine from being rushed through the approval process.

Weller also studied the Coxsackieviruses, named after the town in New York where they were first discovered. Coxsackieviruses are closely related to polioviruses, and Weller documented that when both infections coexist, the Coxsackievirus can make the polio less severe. Coxsackie is the cause of the common childhood illness hand, foot and mouth disease, so called because it causes red spots and blisters on the palms and soles and in the mouth. Coxsackie infection can cause anything from the common cold to serious meningitis to—rarely— weakness or paralysis, which does not become permanent as in polio. Weller also isolated the virus that causes chicken pox (*Varicella*), and demonstrated that it is the same virus that is found in the blisters of shingles (*Zoster*), so it is properly called the *Varicella-Zoster* virus. Later researchers proved that a shingles outbreak is a reactivation of the chicken pox virus that had been dormant in nerve root cells since a childhood chicken pox infection. Weller was also the first to isolate *Rubella* virus, the cause of German measles.

Thomas Weller died in his sleep at the age of ninety-three in 2008.

———

The third Nobel winner in 1954 was Frederick Chapman Robbins, born in Auburn, Alabama, and raised in Columbia, Missouri, with part of his education taking place in France, due to his father's academic appointments. Robbins earned his medical degree at Harvard in 1940 and then served in the army, where he engaged in research on infectious diseases, emerging from the war years at the rank of major.

---

18 T. H. Weller, *Growing Pathogens in Tissue Culture: Fifty Years in Academic Tropical Medicine, Pediatrics, and Virology* (Canton, MA: Science History Publications, 2003).

Robbins joined Enders's group in Boston at Children's Hospital as a research fellow in January 1948, just in time to contribute to the detailed laboratory work that made him the youngest of the Nobel-winning trio. After studies on mumps, measles, and polio, Robbins held posts at City Hospital in Cleveland and was named dean of what is now Case Western Reserve University. From there, he became the first president of the Institute of Medicine (IOM), a nonprofit founded in 1970 as a component of the National Academy of Sciences.

Among many other contributions while at the IOM, Robbins coordinated a study of Reye's syndrome, which is the sudden deterioration in children with viral infections. A few days into the viral illness, the signs and symptoms of Reye's syndrome start with persistent vomiting then progress to unusual sleepiness, disorientation, and confusion, eventually leading to delirium, seizures, and loss of consciousness. Robbins's study was stopped before planned when it quickly became apparent that the children receiving aspirin for viral symptoms were the ones who got Reye's syndrome. Although Reye's syndrome is rare, parents are advised not to use aspirin in children with viral infections.

Robbins finished his career at Case Western as professor emeritus, in which role he developed working partnerships between the university and Uganda to study HIV and tuberculosis. Frederick Robbins passed away at the age of eighty-seven in 2003.

5

Like the Priest at a Wedding

Hugo Theorell won the 1955 Nobel Prize for his work on enzymes, for which he gave this simple definition: "A catalyst [enzyme] is a chemical substance which helps a chemical reaction without taking part in the final production, like the priest at a wedding."[1]

Enzymes are reusable chemicals that assist all of the biological reactions in the body, often by speeding them up. Theorell's scientific research was only a portion of his diverse activities, which spanned a brief stint at scientific spying, inventing a puzzle machine, playing violin at a professional level, joining the water-fluoridation debate, international traveling, cycling, sailing, and gardening.

Theorell was born in 1903 in Sweden and was paralyzed by polio at the age of three. During the worst of his illness, the young Hugo had an amazingly good attitude and was reported to have said, "I can move my eyes at any rate."[2] The disease left him with a bad leg, and although he was excluded from playing ice hockey, he made up for it with upper body strength and even adapted a bicycle for himself. At the age of nine, he underwent a thigh-muscle transplant to shore

1   Henning Theorell, "Hugo Theorell, My Father," NobelPrize.org, nobelprize.org/prizes/medicine/1955/theorell/article/.

2   As quoted in Henning Theorell, "Hugo Theorell, My Father," NobelPrize.org, nobelprize.org/prizes/medicine/1955/theorell/article/.

up his leg weakness, which served him well until he fractured his leg in a car accident at the age of sixty-two, after which he required a leg brace. During his recovery from the muscle transplant, Theorell took up violin, studying under Sweden's best, which eventually required a regular commute to Stockholm. He financed this by inventing and selling a jigsaw machine that produced a complete jigsaw puzzle in ten to twenty seconds.

Theorell was a top student and was encouraged by a friend of the family, Swedish mathematician Gösta Mittag-Leffler, to apply for a position in the physics lab at the University of Berlin. It was 1922, and it seems that Mittag-Leffler was concerned about the Germans' work toward developing a nuclear bomb; he wanted the young Theorell to snoop around and see what he could find out. Within two months, Theorell and the university realized that he did not have a head for quantum theory, and he did not seem to have picked up any useful information on nuclear research.

Theorell returned to Sweden to complete medical schooling and continued to spend a great deal of time developing a mastery of the violin. He eventually became chairman of the Stockholm Concert Association and vice president of the Royal Academy of Music, as well as first violinist in the Mazer Chamber Music Society.

Theorell graduated from the Karolinska Institute in 1930 and then received a Rockefeller Foundation scholarship to study with Otto Warburg in Berlin. It was in Warburg's lab that Theorell discovered how a particular enzyme reaction had two parts, one derived from vitamin B_2 (riboflavin phosphate) and the other part a protein enzyme. Neither could work alone; a coupling was required in order for them to go into action, and thus they were *coenzymes*.[3] Most but not all coenzymes are from vitamins and, like enzymes, can be reused and recycled without changing reaction rate or effectiveness. Since then, scores of coenzymes have been discovered.

---

3   H. Theorell, "Keilin's Cytochrome c and the Respiratory Mechanism of Warburg and Christian," *Nature* 138 (1936): 687.

Popular Coenzymes

The "co" in the popular heart supplement coQ10 is shorthand for *coenzyme*. Vitamin C, vitamin K, and many of the B vitamins also function as coenzymes.

Theorell became internationally known for his studies of enzymes involved in cell respiration, the process by which cells exchange oxygen (literally, cellular breathing). Upon return to Sweden, Theorell became a professor at Uppsala University. A few years after winning the Nobel Prize, Theorell became professor and the head of the Nobel Institute of Biochemistry at the Karolinska Institute.

Theorell played a small part in the Swedish deliberations over the fluoridation of public water supplies, which reflected heated debates on this issue in many countries. In present-day discourse found on the internet, he is inaccurately called either anti or pro fluoridation. The facts are these: a public water fluoridation experiment was conducted in the city of Norrköping, Sweden, for several years in the 1950s, where the drinking water of one-third of the city's districts was fluoridated and the rate of tooth decay there was compared to that of the residents in the rest of the city. Fluoridation was reported as successful in reducing the incidence of tooth decay, with no obvious adverse effects on the population. After a few years, program advocates wanted to expand the experiment to the whole country. Theorell was concerned that the program needed to be studied some more, which was later characterized as an "antifluoridation" position. However, he was in agreement with a number of other physicians whose official opinions were sought as consultants to Sweden's Royal Board of Medicine. The board denied the request to expand the study. Norrköping researchers then lobbied to at least continue the local experiment for another decade, and Theorell was tapped to serve on the council evaluating this request. He and most of the experts on the council recommended the local study be continued.[4] A similar debate was occurring in the US, and the Public Health Task Force there bolstered its profluoridation stance by inaccurately characterizing Theorell's recommendation for more study as "his support of fluoridation." Theorell described what happened next: "The experiment in Norrköping, however, had to be discontinued because of local opinion against fluoridation—a vivid example of how difficult and fragile the question was."[5]

The arguments in favor of fluoridation were based on the prevention of cavities in the absence of obvious signs of harm to the rest of the body. However, in a Swedish dental journal, even the study's advocates conceded, "It may come with negative side effects, hitherto unknown due to a lack of studies generally, and studies targeting those who might be overly sensitive or who consume abnormal amounts of water."[6] Arguments against were in part scientific, especially that fluoride can be a poison to the body's enzymes, with the potential to disrupt multiple body systems in gradual and subtle ways that would tend to be attributed to aging and not recognized as fluoride toxicity. Another concern was that general fluoridation would target anyone, even people with no teeth. In response to the fluoride advocates likening it to vaccination, the opposition pointed out that caries is not a contagious disease, so the individual should bear no responsibility to society for treating his own teeth. An entirely separate and more passionate argument was that fluoridation of public water systems was essentially a forcibly administered form of mass medication, not easily avoided by those who would wish to do so.

---

4   H. Theorell, "Yttranden Av Kungl. Med.-Styrelsens Vetenskapliga Råd Ang. Kariesprofylax Genom Fluor" [Opinions of king. med.-board scientific council, caries prophylaxis through fluorine], *Svensk Tandläkare-Tidskrift* 51, no. 6 (1958): 414–17.

5   Division of Public Health, "Fluoridation is not Banned in Sweden," Department of Health, Education, and Welfare, Public Health Service, National Institutes of Health, January 1970.

6   "Fluor Som Medel Mot Tandröta. Utredning Verkställd Av Expertkommitté Genom Medicinalstyrelsens Försorg," *Svensk Tandläkare-Tidskrift* 47, no. 1 (1954): 1–28.

Swedish fluoridation advocates pushed for a new law. The Water Fluoridation Act was passed in 1962 and allowed localities to apply for a permit to add fluoride their water. Due to years-long delays in specifying the details of such permits, the Norrköping trials did not continue. A few cities did receive permits, but the compliance requirements were so onerous that they never got around to adding fluoride. In 1971, the Swedish parliament voted to repeal the act, and Sweden's drinking water remains unfluoridated today. According to the Fluoride Action Network, in 2020, only eleven countries in the world had more than 50 percent of their population drinking fluoridated water: Australia (80%), Brunei (95%), Chile (70%), Guyana (62%), Hong Kong (100%), the Irish Republic (73%), Israel (70%), Malaysia (75%), New Zealand (62%), Singapore (100%), and the United States (64%). World Health Organization data reported in 2012 demonstrated there is no difference in tooth decay between Western nations that fluoridate their water and those that do not.[7]

Theorell's mentor, Einar Hammarsten, had submitted nominations for him to the Nobel Committee twice in the 1930s for chemistry and again in the 1950s. Theorell was nominated once for medicine in 1945 and had twenty nominations in the 1950s, mostly for chemistry and two for medicine. This reflects the tremendous degree of crossover between the two fields, as there is no prize in biochemistry, a specialty that was just emerging when Alfred Nobel wrote his will in the latter half of the nineteenth century.

Theorell was held in high esteem by his local and international colleagues, as exemplified in the compliments expressed by Professor Hammarsten at the Nobel Prize Award Ceremony: "A fertile imagina tion. An undeviating and critical accuracy. An astonishing technical

---

7   Fluoride Action Network, "Tooth Decay Trends in Fluoridated vs. Non-fluoridated Countries (WHO Data)," fluoridealert.org/articles/50-reasons/who_data01/, August 7, 2012, presenting data from World Health Organization Collaborating Centre for Education, Training, and Research in Oral Health, Malmö University, mah.se/CAPP.

skill. All scientists possess some of these attributes. Very few have all. You are one of these few."[8]

Hugo Theorell died in 1982 at the age of seventy-nine.

---

8   E. Hammarsten, Award Ceremony Speech, 1955, NobelPrize.org, nobelprize.org/prizes/medicine/1955/ceremony-speech/.

6

Self-Experimentation

The 1956 Nobel Prize was shared by Werner Forssmann, André Cournand, and Dickinson Richards for their discoveries concerning heart catheterization and diseases of the circulatory system.

Werner Forssmann was a student at the medical school of the University of Berlin, where he had experimented on himself for his graduate thesis. He and some other students were building on the work done by previous Nobel Prize winners who'd found that eating liver treated a type of anemia.[1] Forssmann wanted to see if liver also enhanced blood health, so he drank a liter of concentrated liver broth daily. It was later determined that it was primarily the vitamin B_{12} in liver that was good for the blood. Self-experimentation was not uncommon in those days, but after medical school, Forssmann took it so far that he horrified and outraged many of his professional colleagues.

Forssmann was twenty-five years old and working in a village hospital to gain clinical experience in surgery when he conducted what would become his Nobel Prize–winning experiment. He was dissatisfied with the conventional methods of getting information about a sick patient's heart and lungs by the exam maneuvers of tapping the chest, listening with a stethoscope, taking a chest X-ray, and doing a heart

---

1   See *Boneheads & Brainiacs*.

tracing (EKG), which still left a lot of inaccuracy and uncertainty as to the exact diagnosis. He'd read about animal experiments that inserted a catheter (tubing) through the large jugular vein in the neck and advanced it to reach the heart chamber—in fact, these heart "caths" had been done in veterinary research for the preceding seventy years.

Forssmann thought the jugular approach would be too messy in humans and saw no reason that a catheter could not be safely introduced through a vein at the elbow, a location that was traditionally used to insert intravenous tubing for giving fluids. He proposed to his clinical supervisor and head of the hospital, Dr. Schneider, that he conduct the experiment on a human—namely, himself. Schneider replied with an unequivocal no, among other reasons, because the procedure had never been tried in a major teaching hospital. Furthermore, Schneider asked, what would he report to Forssmann's mother if things went badly? Forssmann, who was just one year out of medical school, decided to do it anyway.

He had to resort to recruiting an accomplice in order to obtain access to sterile equipment for his illicit experiment. He approached Gerda Ditzen, a nurse. She thought the good doctor should not conduct the experiment on himself and instead offered herself as the subject of the experiment. Forssmann strapped her down onto a gurney, telling her it was a necessary part of the procedure, then pretended to fuss with her arm. He stepped just out of her field of vision and injected anesthesia into the skin of his own arm, then he made a cut into his vein and pushed in a 3.2 mm (0.125 inch) catheter normally used for draining the bladder. He advanced it about a foot before undoing the nurse's restraints, revealing his deception, and getting her assistance to walk down to the basement X-ray room. The X-ray was a fluoroscope, meaning that instead of taking a snapshot, the X-rays continue as you are watching. An assistant held a mirror to the fluoroscope screen so that Forssmann could see that the catheter tip was only up to the level of his shoulder; while watching in the mirror, he advanced the catheter until it came to rest in the right upper chamber of his beating heart. A colleague came rushing in and approached to pull the catheter out. "I had to give him a few kicks on the shin to calm him down," Forssmann recalled.[2] The final X-ray picture proved his accomplishment.

Forssmann was immediately called to Schneider's office, who reprimanded him but followed with congratulations and a handshake for his daring feat. He urged Forssmann to submit a detailed report of the procedure to a medical journal but thought that the purpose of finding out more about heart diagnoses would not be sufficient to justify the radical approach. Schneider gave Forssmann permission to use the procedure to inject medication into the heart chamber of a hospital patient currently dying of infection after an abortion. The heart catheterization worked, but as expected, the patient died anyway. Forssmann wrote up the two case reports with an emphasis on using the procedure to give medicines into the heart.

Forssmann's report was accepted for publication by a major German medical journal, but due to the usual editorial delays, it was not published until the next year.[3] When he gave a short presentation on it at a medical meeting, he was greeted by shocked murmuring, stamping, and even laughter. His uncle, also a physician, pulled the young Forssmann aside and said, "Don't worry. These oafs just don't understand what you're doing. You'll see. You'll get the Nobel Prize for it yet."[4]

Emphasizing the young Forssmann's genius, Schneider used his professional connections to assist Forssmann in getting a training appointment at the prestigious Charité Hospital in Berlin. When the paper was finally published, reactions ranged from outrage to derision. He was called a stunt man and criticized for being unethical. The heart had been considered sacred anatomy, forbidden territory for the irreverent poking and prodding of mere human doctors, and Forssmann

---

2   E. Forssmann, *Experiments on Myself*, (New York: St. Martin's Press, 1976).

3   W. Forssmann, "Die sondierung der rechten herzens" [Probing the right heart], *Klinische Wochenschrift* 8 (1929): 2085–87.

4   E. Forssmann, *Experiments on Myself*, (New York: St. Martin's Press, 1976).

realized he had breached a taboo. He had conducted his bold experiment against the oppressive attitudes of the German medical establishment, which relied on tradition, strongly discouraged innovation and originality, and above all, staunchly disapproved of junior doctors breaking the rules. In short, independent thinking was a dangerous threat to the entrenched status quo. In addition, another researcher claimed he'd done the same procedure years earlier, but since he had not published a report, it could not be proven that Forssmann had "stolen priority."

The senior surgeon at Charité Hospital didn't want the controversy to be attached to his name or institution and was also concerned about the claim that Forssmann had stolen priority. The doctor made fun of Forssmann's goal to become a medical lecturer, telling him, "You might lecture in a circus about your little tricks, but never in a respectable German university!" and relieved Forssmann of his duties.[5]

Nobel Prize Winners Who Experimented on Themselves

Niels Finsen won the 1903 Nobel Prize in Medicine for light therapy, which he had tested on himself for his own chronic medical problems. Elie Mechnikoff was a cowinner of the Nobel in 1908, and in a bout of severe depression he injected himself with serum from a patient with relapsing fever. Frederick Banting won the Nobel in 1923 for insulin research. In WWII, Banting volunteered to work in the Canadian National Wartime Medical Research labs, where he gave himself mustard gas burns in the course of his research on chemical weapons. Charles Nicolle won the 1928 prize for typhus research, and in later years mixed the serum of previously infected persons with some of the typhus bacteria and injected this mixture into himself, from which he had no ill effects. Gerhard Domagk,

---

5   As quoted in E. Forssmann, *Experiments on Myself*, (New York: St. Martin's Press, 1976).

who won in 1939 for the discovery of a sulfa antibiotic precursor, injected himself with sterilized human cancer cells, which were apparently harmless. Max Theiler took his own experimental yellow fever vaccine before winning the prize for it in 1951. Barry Marshall was a cowinner in 2005 for research on the bacteria that contributes to stomach ulcers. He drank a culture of *H. pylori* bacteria and became sick three days later, eventually developing severe inflammation due to the growth of the bacteria in his stomach.

A survey of medical-research institutions on policies on self-experimentation received thirty-seven respondents from the US, Canada, Central and South America, the UK, and the EU. Twenty-five of them had no policy addressing self-experimentation. The most common institutional attitude was that when the subject and investigator are the same person, there is no justification for intervening.

Back at the village hospital, Forssmann got permission to conduct more experiments—on animals and himself and some patients. He ultimately performed a total of nine cardiac catheterizations on himself. He refined the procedure by injecting dye so that the blood vessels would show plainly on X-rays as the catheter advanced. But he continued to be hounded by critics, was turned down for research grants, and finally became thoroughly discouraged from pursuing research. He continued to train as a surgeon and worked in the field of urology. In 1933, Forssmann married Dr. Elsbet Engel, who also became a urologist. They had to transfer to other hospitals because spouses were not allowed to be on staff together.

In the early 1930s, Forssmann registered with the National Socialist (Nazi) Party and remained a card-carrying member until its dissolution after the war. Forssmann wrote with enthusiasm in his 1976 autobiography of being raised in the strong Prussian tradition of honoring the fatherland and being proud of German expansionism. He described the intense propaganda campaign that instilled fear of an internal communist takeover, understandably—he said—driving the average patriot to fervently support Hitler. He noticed the effect on German medicine, including the gradual removal of references in medical papers to non-German researchers or accomplishments: everything scientifically worthy eventually became solely German. German Jewish doctors began to emigrate or just disappear. Medical lecturers openly espoused Nazi propaganda and some fanatics even declared their hospitals to be "Nazi institutions."

There were a few seminal events that sharply raised Forssmann's awareness of the brutality of the Nazi regime, including the Röhm Purge on June 30, 1934, when the elite Schutzstaffel (SS) under the direction of Heinrich Himmler were sent to round up and execute members of the largest wing of the party, the paramilitary Sturmabteilung storm-troopers (SA), including high-up brass of the army and members of the upper classes considered intellectual, critical, or possible spies. Around this time, Forssmann confessed, it became apparent what was really happening to the Jews. Forssmann explained that anyone who opposed the regime would be professionally ruined and impoverished, and so he dutifully displayed his Nazi badge on his hospital coat and at social events.

Forssmann was drafted into the military reserves in 1935 and underwent periodic field training. Meanwhile he got an appointment at a hospital in Dresden where forced sterilizations were being car-ried out on mental patients at the rate of six to eight per day as part of the Nazi eugenics program. There were an estimated five thou-sand forced sterilizations being carried out per month nationwide. Forssmann claimed that although he was an experienced urologist by this time, he was still not a "full surgeon," and so he was excluded from doing the operations. However, he defended the participation of the senior surgeon, Dr. Albert Fromme: "Mental patients were sent to us from the Lobtauer Institute. They'd been thoroughly examined by psychiatrists, and the indications clearly established. I don't believe Fromme would have carried out a single sterilization if the symptoms were at all doubtful."[6] Obviously Forssmann was then and remained an advocate of sterilization of the mentally ill. Fromme and countless other doctors were never named in the Nuremburg trials, and in 1954 Fromme became rector magnificus (president) of the Dresden Medical Academy.

Forssmann was mobilized in the war, and he served in Norway and on the front lines of Russia as well as Germany, eventually achieving the rank of major. Forssmann was required to treat critical wounds in so called "homers"—German recruits who attempted to inflict nonfatal injuries on themselves in futile attempts at getting medically discharged; when sufficiently recovered, they were court-martialed and executed for their crime. At one time, Forssmann was required to witness guillotine executions of conscientious objectors, his job being to certify that the victims were indeed dead as their bodies lie in crude coffins, the severed heads having fallen between their legs.

When Forssmann was assigned to the Brandenburg reserve military hospital, located at the site of a former civilian mental hospital, it still contained a number of mental patients. He was approached with the suggestion of joining the SS so that he could continue his heart cath experiments there on the abundant supply of research subjects. He found this "tempting" but "unacceptable" and tactfully turned it down.[7] He inquired around the hospital staff if the rumors of mass murder of mental patients were really true, and in answer he was taken on a tour of the sanatorium graveyard. There he read fifty-five headstones all engraved with the same date of death—all deaths allegedly from appendicitis. Forssmann spent the last months of the war in a POW camp run by Americans. He walked home in August 1945.

After the war, Forssmann had a hard time finding employment because of his former Nazi Party affiliation, which resulted in a three-year ban on resuming the practice of medicine. He ignored suggestions to join a democratic organization to remedy his social

6   E. Forssmann, *Experiments on Myself*, (New York: St. Martin's Press, 1976).

7   E. Forssmann, *Experiments on Myself*, (New York: St. Martin's Press, 1976).

"credentials," instead establishing himself as a country doctor in a village in the Black Forest, although some reports say he spent three years as a lumberjack while his wife's medical practice supported the family.

Eventually Forssmann discovered that researchers in America had published a paper in 1941 describing improvements in the human cardiac catheterization that he had pioneered on himself. The paper, by André Cournand and Hilmert Ranges, opened with the line, "Forssmann first used catheterization of the right heart on himself." [8] Forssmann gradually began appearing at medical meetings and accepting speaking opportunities as far away as England. He first met André Cournand in 1951 and met Dickinson Richards at the Nobel ceremonies. Forssmann's nomination for the Nobel was leaked to the press in advance, and he was hounded for a photograph, for there was little known about this obscure German doctor who had done his own heart catheterization.

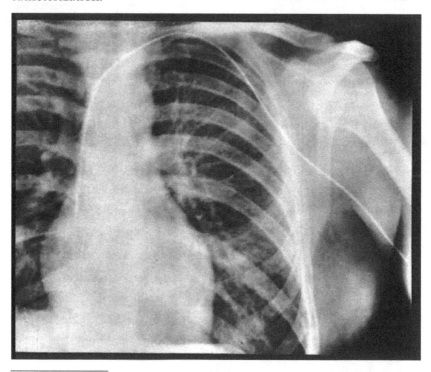

---

8   A. Cournand and H. Ranges, "Catheterization of the Right Auricle in Man," *Proceedings of the Society for Experimental Biology and Medicine* 46 (1941): 462–66.

After winning the prize, his notoriety was a burden, and some questioned if a former Nazi should be so honored. At one point, he received an unsolicited visit from two men on a recruiting mission for the German Party, the postwar neo-Nazi party that would become the National Democratic Party (NDP). He declined to engage them. The Nobel Prize eventually earned him academic respect, and he became chief of surgery at a Düsseldorf hospital. Although he was elected to membership in the American College of Chest Physicians, he was never invited to lecture in the US. Werner Forssmann died after having two heart attacks in 1979 at the age of seventy-three.

———————

André Frédéric Cournand was born in Paris and raised in France. His medical education was interrupted by WWI, in which he worked near the front lines as a battalion surgeon for three and a half years. In 1930, he was accepted to study in the lung-disease section at Bellevue Hospital in New York. This was supposed to be a one-year assignment, but in the midst of it, Cournand accepted an invitation to work in the lab of Dickinson W. Richards on a long-term research project and stayed on indefinitely, eventually becoming a naturalized US citizen.

In 1936, Cournand and Richards tackled the same problems that Forssmann was trying to solve back in 1929: How to get direct information about the heart and lungs in a safe way? They were aware of Forssmann's experiment, and they found that researchers in France had taken up his method, also using the dye that showed up on X-rays, as Forssmann had pioneered.

Cournand and Richards experimented extensively on animals and cadavers. The two continued to increase the ease, precision, and safety of the procedure, and they developed catheters that measured pressure, blood flow, and oxygenation of the heart and lungs. They had accumulated impressive safety data by the time they suggested conducting the procedure on human subjects in 1940. Their proposal was bolstered by the fact that it was wartime and the US government was anxious to support any research on shock, a common condition in frontline casualties.

Using heart caths on humans, Cournand and Richards developed the concept of the cardiopulmonary system, a unique idea that combined the heart and lungs—previously only considered as separate entities. In addition to collecting a great deal of information on cardiovascular shock, they developed ways to measure the actions of cardiac drugs, which transformed treatment of the condition. They were able to visualize congenital heart malformations and learn about chronic lung diseases.

After his retirement, Cournand studied and wrote on his ideas of the future, endorsing the works of French philosopher Gaston Berger. Cournand advocated Berger's concept of "prospective"—the concept of planning for a specifically desired future rather than just pushing forward into inevitable tomorrows. The idea was to create a future, instead of accepting that we have nothing to do with it in the present. This concept is captured in a Berger quote from *Shaping the Future*, Cournand's treatise on the philosophy: "Tomorrow will not be like yesterday. It will be new and it will depend on us. It is less about discovery than it is about invention."[9] Cournand also studied and wrote extensively on the subject of scientific ethics. In 1976, he published his views as "The Code of the Scientist and Its Relationship to Ethics," identifying seven principles: objectivity, tolerance, recognition of error, recognition of priorities, doubt of certitude, unselfish engagement, and sense of belonging to the scientific community.[10]

On February 19, 1988, Cournand passed away peacefully at the age of ninety-two.[11]

---

9   A. Cournand and M. Levy, eds., *Shaping the Future: Gaston Berger and the Concept of Prospective* (London: Gordon & Breach, 1973).

10 A. Cournand, "The Code of the Scientist and Its Relationship to Ethics," *Science* 198, no. 4318 (1977): 699.

11 E. Weibel, *Andre Frederic Cournand 1895–1988: A Biographical Memoir* (Washington, DC: National Academies Press, 1995).

Dickinson W. Richards was born in 1896 in New Jersey. Like Cournand, his medical studies were interrupted by WWI, during which he served as an artillery officer in France. Richards's long service at Bellevue Hospital in New York was a hidden benefit when it came time to take the leap from animals and cadavers to doing heart caths in live humans. According to Richards's brother-in-law Dr. Richard Riley, who worked in Richards's lab, "The early catheterisations [*sic*] were done at Bellevue, the other end of town from CUCPS [Columbia University College of Physicians and Surgeons], where fewer questions were asked."[12]

Bellevue is the oldest public hospital in America. It has always been a full-service hospital, but its notorious history of involuntary psychiatric commitment became so well known that the name became a negative and slightly frightening slang term for a psychiatric hospital. At the same time that Richards and Cournand were performing heart caths at there, the psychiatrist Lauretta Bender was in the mental ward at Bellevue putting children into comas by injecting massive doses of insulin to drastically lower their blood sugar. Bender then moved on to electroshock,[13] especially targeting African American children, whom she considered "best characterized by their 'capacity for laziness' and 'ability to dance,' both of which, she said, were features of the 'specific brain impulses' of African-Americans."[14] She used children at Bellevue to document the stupor-inducing effects of Thorazine (brand name of chlorpromazine, the first "chemical straitjacket") and then LSD, giving daily doses to children for two years or longer.

---

12 M. Dickinson Chamberlin, "Dickinson W. Richards, MD: Through a Grand-Daughter's Eyes," *Coronary Artery Disease* 12, no. 1(2001): 79–82.

13 L. Bender, "One Hundred Cases of Childhood Schizophrenia Treated with Electric Shock," Transactions of the American Neurological Association (72nd Annual Meeting), July 1947.

14 T. Müller, "Dr. Lauretta Bender: Yet Another Acclaimed Psychiatric Quack," Winter Watch, November 3, 2020, winterwatch.net/2020/11/dr-lauretta-bender-yet -another-acclaimed-psychiatric-quack/.

Similar to Cournand expounding on scientific ethics, Richards spoke and wrote on the problems of what he called the medical priesthood. He criticized the worship of conformity, especially in technique and language, for example, as expressed in the saying "We do a surgical procedure in that way because it's always been done that way." The human heart had been taboo before the first heart cath by Forssmann for no other reason than it was just not done! Richards showed how this mindset leads to conformity of thought and restrictions against thinking outside the box. Richards observed that medical science also tended to worship popularity, which he described as "taking the easy ride on already surging tide; to plant more seed in an already well-ploughed field" and discouraging creative scientific advances. He condemned laxness, which he said leads to "the disregard of small errors, of deviations, of the unexpected response: the easy worship of the smooth curve."[15] The many retracted and falsified medical journal articles are abundant evidence that laxness, to put it mildly, is still a significant problem today. In 2018, a searchable database was established to help doctors to identify which published medical journal articles have not held up to scrutiny.[16] Finally, Richards attacked the role of fear in the medical priesthood, specifically "fear of speculation; the overprotective fear of being wrong." This fear is what got Forssmann fired from Charité Hospital. Richards said this fear leads one to be "forgetful of the curious and wayward dialectic of science, whereby a well-constructed theory even if it is wrong, can bring a signal advance."[17]

Dr. Richards was mandatorily retired at age sixty-five, whereupon he was promptly demoted to a small office under a set of stairs.

---

15 D. W. Richards, *Medical Priesthoods and Other Essays* (Hartford: Connecticut Printers, 1970).

16 J. Brainard and J. You, "What a Massive Database of Retracted Papers Reveals about Science Publishing's 'Death Penalty,'" *Science*, October 25, 2018, sciencemag.org/news/2018/10/what-massive-database-retracted-papers-reveals-about-science-publishing-s-death-penalty.

17 D. W. Richards, *Medical Priesthoods and Other Essays* (Hartford: Connecticut Printers, 1970).

He served as a senior editor for *The Merck Manual of Diagnosis and Therapy*, a veritable bible for medical students around the world before the advent of handheld electronic reference devices.

The right heart cath techniques pioneered by Forssmann and perfected by Richards and Cournand are used today with little change. The caths done by the Nobel winners were focused on the management of critically ill patients, while the most common reason for a heart cath today is a left heart cath to inject the coronary arteries to look for blockages. A study in 2010 found that up to one-third of the time, the left heart caths done in America do not find any heart disease, suggesting massive overuse of the procedure.[18]

---

18 M. R. Patel et al., "Low Diagnostic Yield of Elective Coronary Angiography," *New England Journal of Medicine* 362 (2010): 886–95.

7

Drugs

Daniel Bovet won the 1957 Nobel Prize for his discoveries of how certain substances acted on the body, leading to the development of new drugs.

Bovet was born and educated in Switzerland, where his father taught him Esperanto as a native language, but he was equally fluent in Italian, French, and English. He graduated from the University of Geneva at the age of twenty and received a doctorate degree there two years later. He researched at the Pasteur Institute in Paris for eight years, then in 1947, he settled in Italy.

Esperanto

Esperanto is a constructed language, made up rather than naturally developed. It was invented by Polish ophthalmologist L. L. Zamenhof in 1887 with the goal of creating a universal second language to foster world peace and international understanding. Zamenhof listed nine hundred word roots; these could be expanded into tens of thousands of words.[1] Esperanto was suppressed in Nazi Germany, imperialist Japan, Francoist Spain up until the 1950s, and in the

---

1   L. Zamenhof, *An Attempt towards an International Language* (New York: H. Holt, 1889).

Soviet Union under Stalin. Native speakers, as was Bovet, are currently estimated to number about two thousand worldwide, with about two million using Esperanto as a second language. Speakers are called Esperantists.

Bovet's wife, Filomena Nitti (sometimes Bovet-Nitti), was his constant research partner, among many other notable biochemists and pharmacologists who passed through his labs. In his Nobel lecture, Bovet credited his wife's participation and he mentioned her contributions twice more in his Nobel banquet speech.[2] Despite her involvement in years of research, there is very little public information about Filomena Nitti except that she was the daughter of Francesco Saverio Nitti, the prime minister of Italy in 1919 and 1920, who was exiled during the Fascist era.

Bovet studied the actions of curare, an extract from the refined sap of various plants whose first known use was to make poison arrow tips. In the rainforest, the curare-rich sap of the climbing shrub *Strychnos toxifera* is used on the points of darts shot from blowguns. It completely paralyzes an animal in less than a minute. This research led Bovet to study succinylcholine—*sux* for short—a curare-like man made drug known to cause total muscle paralysis.

Bovet's research provided additional information on how sux works. When an electric signal travels down a nerve, it ends at the junction of nerve and muscle fiber. The nerve impulse signals release of a chemical that crosses the junction to activate muscle fibers to fire. The chemical transmits the impulse, so to speak; thus, it is called a neurotransmitter. This action occurs millions of times per hour, making it possible to open our eyelids, lift a finger, and do biceps curls in the gym. Sux blocks the action of the neurotransmitter, preventing muscles from firing. It has a rapid onset and doesn't last too long before

---

2   D. Bovet, "The Relationships between Isosterism and Competitive Phenomena in the Field of Drug Therapy of the Autonomic Nervous System and That of the Neuromuscular Transmission," December 11, 1957, NobelPrize.org, nobelprize. org/prizes/medicine/1957/bovet/lecture/.

the body breaks it down, but while the drug is active, the patient cannot move a muscle—even the muscles of breathing. Some researchers self-experimented with sux, as reported in a 1952 account of the injection of a very small dose: "After 30 seconds: onset of muscle weakness, double vision, ptosis [drooping eyelids], muscle twitching of eyelids, jaw muscles, arms, legs and buttocks. After 60 seconds: partial paralysis of respiratory muscles, difficulties in speaking, head and arms cannot be raised."[3]

Dr. Bovet's research helped sux to become a safely administered medication in anesthesia, that is, as long as a patient is knocked out first. Getting paralyzed while still completely awake can be terrifying, as related by a surgical colleague who broke his leg in a motorcycle accident. He was rushed to the operating room and given sux to induce paralysis. He was completely alert and could hear and feel the operation proceeding but could not twitch an eyelid or move a finger to let anyone know he was still awake. It was a harrowing experience. Then the patient heard the anesthesiologist react to his fast pulse, cursing upon realizing they had forgotten to give a sedative to put the patient under before paralysis.

Murder by Sux

Sux gained notoriety for a short time as the perfect murder weapon because it could be injected and cause death rapidly by paralyzing the breathing muscles and by the time the ambulance arrived, sux would already be broken down in the body. The case of Dr. Coppolino illustrates the medicolegal difficulties. Carl Anthony Coppolino, a New Jersey anesthesiologist, was on the disability rolls at the age of thirty because of a heart condition, living off his disability benefits, some writing royalties, and his physician-wife's income. Coppolino began a torrid affair with Marjorie Farber, the

---

3   O. Mayrhofer, "Self-Experiments with Succinylcholine Chloride: A New Ultra-Short-Acting Muscle Relaxant," *British Medical Journal* 1, no. 4772 (1952): 1332–34.

wife of retired army officer Lieutenant Colonel William E. Farber. In 1966, Mr. Farber was found dead of an apparent heart attack. The affair with Mrs. Farber soon fizzled, and Coppolino moved down to Florida with his wife. There he took up with another woman, Mary Gibson. In 1967, Coppolino's wife was found dead of an apparent heart attack at the age of thirty. Coppolino married Gibson less than six weeks after his wife's death. At this point, Coppolino's spurned former mistress Mrs. Farber revealed that she had been hypnotized into attempting murder on her husband by injecting him with sux, then stood by while Coppolino completed the act by suffocating Lieutenant Colonel Farber with a pillow. This led to murder trials in two states and exhumations of bodies. The only things detected in the urine of the deceased were the breakdown products of sux—succinic acid and choline—which didn't necessarily prove that succinylcholine had been given. F. Lee Bailey was Coppolino's defense attorney at both trials, and he won his client a not-guilty verdict for Farber's death in New Jersey, but Coppolino was convicted of second-degree murder for his wife's death in Florida. He was paroled after serving twelve and half years.[4] Detection methods have improved, and now sux and related drugs can be detected in the blood and urine eight to twenty-four hours after administration.[5] Beware of a highly suspicious death when someone close by has access to succinylcholine!

Bovet is best known for his work on antihistamines, commonly used to treat allergies.[6] Histamines are manufactured in the body but can also come from certain foods, including alcohol; citrus fruit; packaged meat; aged cheese; fermented foods, like sauerkraut; and legumes,

---

4 P. Holmes, *The Trials of Dr. Coppolino* (New York: New American Library, 1968).

5 U. Kuepper et al., "Degradation and Elimination of Succinylcholine and Succinylmonocholine and Definition of Their Respective Detection Windows in Blood and Urine for Forensic Purposes," *International Journal of Legal Medicine* 126, no. 2 (2012): 259–69.

6 D. Bovet, "Introduction to Antihistamine Agents and Antergan Derivative," *Annals of the New York Academy of Sciences* 50, no. 9 (1950): 1089–126.

such as kidney beans, chickpeas, and peanuts. The histamines in foods are usually not a problem unless the person has a genetic composition that makes them unable to metabolize histamine as fast as it is being made or consumed, in which case excess histamines build up and cause symptoms. The actions of histamines depend on which type of histamine receptor they interact with—the receptor is kind of like a custom docking station on the cell. One type of histamine receptor in the brain helps with alertness and others are responsible for motion sickness, but when an antihistamine is taken for the latter condition it will block the alertness receptor, too, often causing sleepiness. Excessive activation of histamine receptors cause the typical signs of an allergic reaction, causing dilation of the blood vessels, hives, and spasms of the airways. Antihistamine drugs block these receptors, successfully suppressing allergic reactions, but at the expense of making the patient drowsy because they are also blocking the alertness histamine receptors in the brain. Another type of histamine receptor regulates acid in the stomach, and antihistamines used to be the most common treatment for acid reflux and ulcers until newer drugs with fewer side effects were discovered.

Bovet also contributed to the discovery of antibiotics, anti-Parkinson's drugs, and medications to treat thyroid problems. Later in life, he investigated psychiatric drugs.

Bovet was hardly known outside of academic circles until he won the Nobel Prize in 1957. Unlike many productive scientists, Bovet never sought a patent in his name for his work. The nominations are kept secret for fifty years, but today we know that Bovet had been nominated only twice (in 1951 and 1953). He was lying in bed with the flu when he was notified of the win by a call from the Swedish ambassador. At the banquet speech, Bovet admitted to being perplexed that he was chosen and said he searched in vain for what it was in his steady, gradual research that caught the notice of the Nobel Committee:

> The very high distinction which I received from the Karolinska Institute indeed fills me with gratitude, pride

and joy, but also gives rise to some perplexity in me. As a scientist accustomed to establishing logical links between events, I search in vain for the link which has been able to transform what has always been daily life in the laboratory for my wife and for me, the small incidents of experimentation, the meticulous and objective research of each day, the collaboration with my assistants, sometimes the boredom of proofreading, by a wave of the magic wand of the Academy, in the unforgettable and luminous hours that we are living this evening.[7]

Afterward he closely guarded his privacy, but there was one other spot of notoriety, in 1965, when the *New York Times* picked up his comments on nicotine. Bovet had been researching nicotine, a brain neurotransmitter, finding that it enhanced learning in some genetic strains of lab rodents.[8]

TOBACCO CALLED HELP IN LEARNING ;
BIOCHEMIST TELLS OF BENEFIT IN SMALL DOSES OF SMOKE

Special to the *New York Times*, April 18, 1965

ROME, April 17—Tobacco smoke in small quantities "produces a certain positive stimulus in the intellective and learning faculties," according to an Italian biochemist and Nobel Prize–winner of the University of Sassari, Sardinia.[9]

---

7 U. Kuepper et al., "Degradation and Elimination of Succinylcholine and Succinylmonocholine and Definition of Their Respective Detection Windows in Blood and Urine for Forensic Purposes," *International Journal of Legal Medicine* 126, no. 2 (2012): 259–69.

8 D. Bovet and F. Bovet-Nitti, "Action of Nicotine on Conditioned Behavior of Naive and Pretrained Rats," in *Tobacco Alkaloids and Related Compounds*, ed. U. S. von Euler (Oxford: Pergamon Press, 1965), 125–42.

9 "Tobacco Called Help in Learning: Biochemist Tells of Benefit in Small Doses of Smoke," *New York Times*, April 18, 1965.

Since Bovet's initial discoveries, research has continued into the effects of nicotine and nicotinic receptors. It can enhance learning and memory in rats, mice, monkeys, and zebrafish.[10] Some research in humans has shown that nicotine improves cognitive performance (learning and intellectual function) in smokers beyond their baseline levels. This is no reason to adopt the habit, as it also known that cognitive decline tends to occur in middle-aged heavy smokers.[11] Other disadvantages seriously outweigh the risks: According to the Centers for Disease Control, life expectancy for smokers is at least ten years shorter than for nonsmokers, with excess deaths from cancer, lung disease, and heart and blood vessel diseases. Smokeless tobacco is not a great alternative either, as it also causes cancer. Quitting smoking before the age of forty reduces the risk of dying from smoking-related disease by about 90 percent.[12]

Daniel Bovet died of cancer in 1992 at the age of eighty-five. Filomena Nitti, his research partner and wife, died in 1995.

10 D. Eddins et al., "Nicotine Effects on Learning in Zebrafish: The Role of Dopaminergic Systems," *Psychopharmacology* 202 (2009): 103.

11 M. W. Campos et al., "Smoking and Cognition," *Current Drug Abuse Reviews* 9, no. 2 (2016):76–79.

12 "Tobacco Related Mortality," Centers for Disease Control, cdc.gov/tobacco/data_statistics/fact_sheets/health_effects/tobacco_related_mortality/index.htm.

8

From Wahoo to Outer Space

The 1958 Nobel Prize was shared among three winners. Half the prize was split by George Beadle and Edward Tatum for showing that genes control individual steps in metabolism. The other half went to Joshua Lederberg for his related work on how genes function in bacteria.

George Wells Beadle was born to a farming family in Wahoo, Nebraska. After completing agricultural studies at the University of Nebraska, he earned his PhD at Cornell. He had an illustrious scientific career including stints at the California Institute of Technology (Caltech), Harvard, Stanford, and Oxford, and he ultimately became president of the University of Chicago.

Wahoo

Wahoo sounds like a fictional name for hicksville, which is not far from the truth. It's from an Indian word meaning *burning bush*. Another famous native of the 150-year-old Nebraska farming town was Daryl Zanuck. Zanuck started in the movie industry as a writer for the Rin Tin Tin series at Warner Brothers and then founded his own studio called Twentieth Century Pictures, which later merged with Fox to become Twentieth Century Fox. Other famous Wahoos include Baseball Hall of Famer Samuel "Wahoo Sam" Crawford, symphonic musician and Pulitzer Prize–winner Dr. Howard Hanson, and Clarence W. "Herk" Anderson, famous for his horse illustrations and children's stories.

George Beadle's wife, Muriel Beadle, was a gifted writer, and the couple cowrote the 1966 book *The Language of Life: An Introduction to the Science of Genetics*. The book is aimed at the nonscientific reader, so it avoids scholarly terminology, and according to one reviewer, "Each scientific word and concept has been pounded down into everyday language."[1] Like so many genetics researchers, Beadle believed that he was cracking the code of life. However, every time a piece of the genetic puzzle was solved, more fundamental questions arose. In *The Language of Life*, the Beadles delve into the deeper question—Who or what made DNA?

They describe that atoms—which became molecules and eventually assembled into DNA—came into being entirely by chance. Consisting only of a proton, bearing one unit of positive electrical charge, and an electron, bearing one unit of negative electrical charge, hydrogen is presumed to have been the first element. Nowadays, hydrogen is considered to have arisen at the time of the big bang that formed the universe. It is theorized that after hundreds of millions of years, some radiation knocked an electron apart to form the next element, helium. Random accidents continued, and eventually, lithium was made from helium, then beryllium, boron, etc., until, at last, all of the chemicals needed to make the building blocks of DNA were accidentally made. Then over many millions more years, so the theory goes, these molecules spontaneously met up and formed stable amino acids—from which proteins were eventually formed. Billions of accidents later, out of a sea of ammonia and methane gas, viruses were formed; then bacteria,

---

1   Attributed to Tribune staff writer Ronald Kotulak, as quoted in the obituary, Kenan Heise, "Muriel Beadle, Free-Lance Writer, Author," *Chicago Tribune*, February 22, 1984.

animals, and man. The Beadles' detailed description of this theory ends with a question: "If all life on earth evolved from hydrogen, where did the hydrogen come from?"—and a supreme creator is implied.

———

Beadle's coresearcher Edward Lawrie Tatum was born in Boulder, Colorado. He attended the University of Chicago and then the University of Wisconsin–Madison, where he received his bachelor's degree and doctorate. Tatum began his collaboration with Beadle at Stanford University. He moved to Yale University, then returned to Stanford, and finally joined the faculty of the Rockefeller Institute.

Tatum and Beadle were at the forefront of a new scientific field that combined biology with chemistry, which would eventually be called biochemistry. In the biology lab, they were regarded as chemists who did not belong; in the chemistry lab they were seen as biologists. This was problematic at academic institutions where biology was still firmly entrenched in the narrow study of the forms of plants and animals, and it created difficult working conditions that contributed to the frequent moves that both made in their early and middle careers.

Before working with Beadle, Tatum and his colleagues postulated that all life forms shared common basic building blocks: amino acids, sugars, lipids, and growth factors (and later nucleic acids). Tatum was coresearcher on a study that demonstrated this was true even for bacteria.

When Tatum joined Beadle at Stanford, they were working with fruit flies to study how the genes controlled appearance. The fruit fly is a short-lived insect that is easy to breed, and it had been used in genetics experiments for decades. The flies were large enough so that changes in their eye colors could be easily seen, but it was terribly inconvenient to maintain glass boxes of swarming flies in the lab. Meanwhile, laboratory methods were getting better, allowing scientists to work on tinier organisms. This led Beadle and Tatum to embrace Tatum's discoveries that the same basic biochemical components seen in a fruit fly would also be seen in a microbe. They abandoned the messiness of fruit flies and instead worked with culture plates of molds. They subjected bread molds to X-rays, causing sections of the mold's genes to mutate. By applying the methods of biochemistry to detect what had changed in the mold, they discovered that the mutations caused changes in specific metabolic pathways. They tracked this, in turn, to changes in specific enzymes involved in those reactions. This confirmed the hypothesis that genes code for enzymes, and that a single gene coded for a single enzyme—what's known as the "one gene, one enzyme" hypothesis. This work, conducted in 1941, won Beadle and Tatum the Nobel Prize in 1958.

Beadle remained interested in the biochemistry of molds, but Tatum went on to explore the use of bacterial cultures for genetics work. It had previously been thought that bacteria simply divide to produce carbon copies of themselves, with identical genes in the subsequent generations. Tatum and his student Joshua Lederberg designed experiments demonstrating that bacterial genes recombine in ways similar to the sexual reproduction that takes place in more complex organisms. It was found that two bacteria make a bridge-like connection and exchange genetic material across it, making the next generation of bacteria genetically different. Because bacteria are even easier and much faster to grow than molds, this work established the use of bacteria to study genes, metabolic reactions, and enzymes, and it was still in use decades later.[2]

Joshua Lederberg was born in New Jersey in 1925. He knew he wanted to be a scientist early on, writing when he was just seven years old, "I would like to be a scientist of mathematics like Einstein. I would study

---

2   J. Lederberg, *Edward Lawrie Tatum 1909–1975: A Biographical Memoir* (Washington, DC: National Academy Of Sciences, 1990).

science and discover a few theories in science."[3] At the age of fifteen, he graduated from New York's elite Stuyvesant High School for math and science. He won a scholarship to Columbia, where he earned his degree in zoology and began medical school. The war interrupted his studies, and after the war he abandoned medical school to work under Tatum at Yale, where they conducted joint research on how bacteria swap genes.[4] Lederberg then accepted a teaching post at the University of Wisconsin, focusing instead on basic microbiology research. There he met and married fellow microbiologist Esther Zimmer. Esther Zimmer Lederberg was coauthor with Joshua Lederberg on seven papers; on four of those her name preceded that of her husband, while on two she was lead author. In his Nobel lecture Joshua Lederberg mentioned her twice and cited six of her papers.

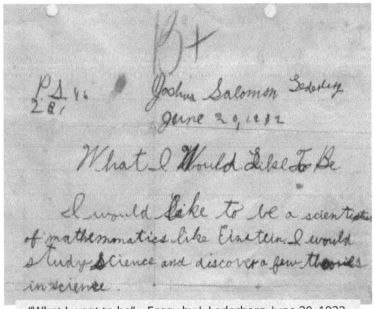

"What I want to be"—Essay by J. Lederberg, June 20, 1932, Joshua Lederberg Papers, Profiles in Science, National Library of Medicine, profiles.nlm.nih.gov/101584906X1.

---

3   Essay by J. Lederberg, June 20, 1932, Joshua Lederberg Papers, Profiles in Science, National Library of Medicine, profiles.nlm.nih.gov/101584906X1.

4   E. L. Tatum and J. Lederberg, "Gene Recombination in the Bacterium Escherichia coli," *Journal of Bacteriology* 53, no. 6 (1947):673–84.

In 1951, together with his student Norton Zinder, Lederberg showed that genetic material could be transferred from one strain of bacterium to another using viral material as an intermediary step.[5] In a process called *transduction*, a virus copies a portion of genetic information from bacterium strain A and transmits it to be incorporated into the genes of bacterium strain B. In the 1950s, together with his wife and other researchers, Lederberg documented a specialized form of transduction that swaps genes to rapidly share antibiotic resistance among different species of bacteria.[6]

The Lederbergs next joined the staff at Stanford, Joshua as the founder and chair of the Genetics Department and Esther in a non-tenured position.[7] At least she was employed. By 2016, women represented just over one-fourth of the professors at Stanford, with a widening disparity at the higher level of full professors, where only 22 percent were women. In 2020, the Stanford Biochemistry Department had only eight women among twenty faculty members plus eight professors emeritus.[8] Joshua Lederberg was notified of winning the Nobel shortly after arriving at Stanford, and he couldn't have been more surprised.

> I knew about Nobel Prizes, and I thought wonderful people had gotten this accomplishment. I thought of it then as being something that might grace the end of your career and had given it absolutely no thought and would have thought it preposterous that I was even being considered for it. This was not modesty. I felt my work was potentially

5   N. Zinder and J. Lederberg, "Genetic Exchange in Salmonella," *Journal of Bacteriology* 64, no. 5 (1952): 679–99.

6   M. L. Morse, E. M. Lederberg, and J. Lederberg, "Transduction in Escherichia Coli K-12," *Genetics* 41, no. 1 (1956): 142–56.

7   R. Ferrell, "Esther Miriam Zimmer Lederberg: Pioneer in Microbial Genetics," in *Women in Microbiology*, ed. R. Whitaker and H. Barton (Washington, DC: ASM Press, 2018).

8   "Faculty," Stanford University Department of Biochemistry, biochemistry.stanford. edu/faculty.

as important as that of some other Prize winners, but I thought I was very young and I thought there was plenty of time to let that work out, so I was quite astonished when it came about.[9]

Lederberg was only thirty-three at the time of his Nobel Prize, and he had an adventurous career after the Nobel. In 1957 when Sputnik was the first spacecraft launched from Earth, Lederberg became concerned with the prospect of microbial contamination. His initial notes on this are titled "Cosmic Microbiology," and discuss the very real possibility that extraterrestrial microbes could hitch a ride to Earth on returning spacecraft, causing catastrophic diseases. In the other direction, microbes carried out to space on man-made objects could confuse the search for extraterrestrial life.[10] Lederberg called this new field of inquiry exobiology (its since been renamed astrobiology). He wrote to the National Academy of Sciences to advise sterilization of equipment prior to launch and quarantine for returning astronauts. When NASA was formed in 1958, Lederberg became a consultant, and his concerns were taken seriously. Lederberg was consulted on the 1975 Viking voyage, and he developed a device for conducting biochemical analysis of soil samples from space. Today the Planetary Protection Center of Excellence at Jet Propulsion Laboratories of the California Institute of Technology has the mission of minimizing the risks Lederberg pointed out:

> To prevent either forward or backward contamination, spacecraft hardware must be sterilized and evaluated for the presence of microorganisms. The hardware must be maintained in cleanroom environments to control for

---

9   Interview with Joshua Lederberg by Barbara Hyde, March 22, 1996, video recording, Profiles in Science, National Library of Medicine, profiles.nlm.nih.gov/101584906X12269.

10 Notes on cosmic microbiology by J. Lederberg, Profiles in Science, National Library of Medicine, profiles.nlm.nih.gov/101584906X13169.

biological contamination, and environmental contamination such as dust particles and moisture. Special air filtering, personnel garments and personnel disciplines, provide and maintain a specified level of cleanliness. Planetary Protection personnel routinely sample the cleaned hardware to verify its cleanliness and ensure that the level of biological contamination (if any) is within specified requirements. Each exercise, from cleaning the hardware, through to determining the level of contamination and taking action to prevent re-contamination, requires the coordinated participation of a myriad of employees such as hardware engineers, quality assurance engineers, and Planetary Protection engineers.[11]

The field of astrobiology is now inclusive of the search for extraterrestrial life, which starts by identifying the characteristics of celestial bodies that could conceivably support life.

In the 1960s, Lederberg collaborated with Stanford's Computer Science Department to develop DENDRAL, an artificial intelligence program with the aim of empowering a computer with scientific reasoning. To this end, DENDRAL was given the specific task of helping chemists identify unknown molecules—conducting chemical analyses to predict molecular structures—with the computer being able to make decisions and solve problems.[12] In a later interview, the scientist coined "Lederberg's principle": "If there's one principle that I'd like to have my name attached to, Lederberg's principle is that machines will become really smart only when 1) they can directly read the literature

---

11 "Mission Implementation," Jet Propulsion Laboratory, California Institute of Technology, planetaryprotection.jpl.nasa.gov/mission-implementation.

12 J. Lederberg, "How DENDRAL Was Conceived and Born" (paper presented at the ACM Symposium on the History of Medical Informatics at the National Library of Medicine, November 5, 1987), profiles.nlm.nih.gov/spotlight/bb/catalog/nlm:nlmuid-101584906X1010-doc.

and 2) spend some time living in the real world where the survival of the fittest is what will determine who's out there."[13]

In 1978, Lederberg became the president of Rockefeller University, serving until 1990, when he became professor emeritus there. He continued to research in the field of DNA and computer learning. In a 1996 interview, he communicated the essential ingredient of scientific creativity, one it seems computers are unlikely to supplant:

> ...the effective scientist—the really creative one—is the person who one minute can be fantasizing with no restraints, imagining all kinds of things—blue sky—and the next can be his or her own harshest self-critic in saying "Look, this is what reality is really like: A, B, C, D, and G are not going to work. Well maybe F survived and that's what's worth considering." But to be able to have a lot of fluency in scanning a wide range of fantastic notions— things that had never been done before—which probably aren't going to work and give them little bit of attention and then have a critical regard for how to discard them. If there's any essential ingredient to scientific creativity, I think that's where it would be.[14]

Lederberg died in 2008 at the age of eighty-two. Tatum, a heavy cigarette smoker, died in 1975 at age sixty-four of heart failure complicated by chronic emphysema. Beadle, a nonsmoker, died in 1989 at eighty-six years old. Beadle had this advice for budding young brainiacs:

---

13 Interview with Joshua Lederberg by Barbara Hyde on the Lederberg Principle, March 22, 1996, video recording, Profiles in Science, National Library of Medicine, profiles.nlm.nih.gov/spotlight/bb/catalog/nlm:nlmuid-101584906X12264-vid.

14 Interview with Joshua Lederberg by Barbara Hyde, March 22, 1996, video recording, Profiles in Science, National Library of Medicine, profiles.nlm.nih.gov/101584906X12269.

"You too can win Nobel Prizes. Study diligently. Respect DNA. Don't smoke. Don't drink. Avoid women and politics. That's my formula."[15]

---

15 G. Beadle, as quoted in D. Pratt, "What Makes a Nobel Laureate?," *Los Angeles Times*, October 9, 2013.

9

The Men (and Women) of DNA

The 1959 Nobel Prize was shared by Severo Ochoa and Arthur Kornberg for their discoveries of enzymes that make genetic material.

Kornberg's prizewinning research focused on DNA (deoxyribo nucleic acid), which is the chemical name of the compound that carries the genetic codes for making, repairing, and controlling life-forms. He put it very simply: "DNA, like a tape recording, carries a message in which there are specific instructions for a job to be done."[1]

Ochoa's prize was for work on RNA (ribonucleic acid), which copies segments of DNA so that the body can follow the coded instructions to make the biochemicals for life.

These discoveries have led to the entire field of genetic engineering, wherein DNA is manipulated to produce desired effects. In agriculture, genetic engineering is used on food plants to make them more productive, nutritious, and resistant to disease. All too often, however, the emphasis in crop science is on increasing production at the expense of nutrition and disease susceptibility. In medicine, genetic engineering plays a significant role in the production of drugs and vaccines as well as in the development of tests for various diseases. An extension

---

1   A. Kornberg, "The Biologic Synthesis of Deoxyribonucleic Acid," December 11, 1959, NobelPrize.org, nobelprize.org/prizes/medicine/1959/kornberg/lecture/.

of genetic engineering is gene therapy, by which healthy genes are inserted directly into a person with malfunctioning genes, as in cases of cystic fibrosis and muscular dystrophy. This promising field is still in its infancy, with gene therapies strictly experimental. In industry, genetic engineering is being used to develop bacteria that can gobble up oil spills, develop biodegradable plastics, speed up the process of metal-ore extraction, and improve the efficiency of fuels.

Severo Ochoa was born in Spain in 1905, the youngest of seven children. After completing medical school in Madrid, Ochoa participated in research in several labs in Spain and Germany, including work at the Kaiser Wilhelm Institute under Nobel Prize–winner Otto Meyerhof. The Spanish Civil War forced his relocation to England to continue research. When WWII erupted, all scientific resources in England were turned to the war effort, but as a foreigner, Ochoa was excluded from this work. This prompted him to move to the United States to join the labs of Gerty and Carl Cori (soon to be winners of the 1947 Nobel Prize). Finally, Ochoa settled at New York University, first in the Department of Medicine then in pharmacology and biochemistry.

Ochoa was active in many areas of biochemistry research, but he focused on enzymes, with an initial concentration on the enzymes involved in basic metabolism. He contributed to discoveries of how vitamins are incorporated into various enzymes involved in the conversion of foodstuffs into energy for the cells. Ochoa supervised laboratories at NYU, assigning his doctoral students and postdoc fellows to research projects.

In 1955 a discovery of a new enzyme was made by Marianne Grunberg-Manago, a postdoctoral researcher in Ochoa's NYU lab since 1953 who had been born in St. Petersburg and educated at the University of Paris. Unlike other enzymes under study, the new enzyme did not cause chemical reactions in the steps of metabolism. Instead, it catalyzed the assembly of amino acid building blocks into large molecules of RNA. This was the first time such an artificial synthesis occurred outside of a living organism.

At first, Ochoa doubted Grunberg-Manago's results, saying, "It was impossible." The postdoc showed Ochoa the lab process, and he quickly verified her findings, which caused him to apologize for his initial negative reaction. Ochoa wanted to name the new substance RNA synthetase (referring to an enzyme that makes RNA), but Grunberg-Manago thought it was more likely an enzyme that also or even primarily degrades RNA, so she wanted to call it phosphorylase. Ochoa conceded, saying, "Marianne, because I like you very much, I will adopt the name you suggested."[2] Subsequent experiments demonstrated that Grunberg-Manago was correct. In her test tube, the enzyme did link together a few of the building blocks of RNA, but it was a highly reversible process and in life it mainly worked in the other direction—to break down RNA. The primary enzyme that builds RNA was later discovered and called RNA polymerase.

Grunberg-Manago's initial presentation on the discovery riveted the attention of biochemists worldwide. What she produced in the test tube is today called messenger RNA, or, mRNA. It is the substance that lines up against DNA to copy a portion of the genetic code. Then another type of RNA directs its translation into proteins, and yet another RNA transfers amino acid building blocks to the cell to be assembled into proteins.

The synthesis of RNA in the lab was recognized as the first step in cracking the genetic code. The breakthrough sharply turned research efforts in all major academic biochemistry labs toward the problems of how to make RNA and DNA. Grunberg-Manago published three papers on these discoveries with her name appearing as first author

2   M. Grunberg-Manago, "Severo Ochoa: 24 September 1905–1 November 1993," *Biographical Memoirs of Fellows of the Royal Society* 43 (1997): 350–65.

and Ochoa's as second or third author.[3] The first author is usually the person who has made the most significant intellectual contribution to the work in terms of designing the study, acquiring and analyzing data from experiments, and writing the manuscript. The first author is also typically the person who does the actual work of the experiment. Yet Ochoa received the Nobel Prize in 1959 primarily for the discovery documented in the papers Grunberg-Manago had been first author of. He was already a world-recognized enzyme researcher, had worked in the labs of Nobel Prize winners, and headed a department at a prestigious university. Ochoa mentioned Grunberg-Manago once in his Nobel lecture, although he cited three of her research papers.

While Marianne Grunberg-Manago made the Nobel-winning discovery, she was not even nominated and so could not be considered by the Nobel Committee. The prize went to the supervisor of her lab. The Nobel organization's web page describing the facts of Severo Ochoa's prize does credit her: "By studying bacteria, Ochoa and Marianne Grunberg-Manago discovered an enzyme in 1955 that can join nucleotides—the building blocks of RNA and DNA—together."[4] Grunberg-Manago returned to France and eventually became the first woman to head the International Union of Biochemistry. She was also the first woman to preside over the French Academy of Sciences.

---

3   M. Grunberg-Manago and S. Ochoa, "Enzymatic Synthesis and Breakdown of Polynucleotides; Polynucleotide Phosphorylase," *Journal of the American Chemical Society* 77, no. 11 (1955): 3165–66; M. Grunberg-Manago, P. Ortiz, and S. Ochoa, "Enzymatic Synthesis of Nucleic Acid-like Polynucleotides," *Science* 11 (1955): 907–10; M. Grunberg-Manago, P. Ortiz, and S. Ochoa "Enzymic Synthesis of Polynucleotides. I. Polynucleotide Phosphorylase of Azotobacter vinelandii," *Biochimica et Biophysica Acta* 20, no. 1 (1956): 269–85.

4   "Severo Ochoa: Facts," December 29, 2020, NobelPrize.org, nobelprize.org/prizes/medicine/1959/ochoa/facts/.

The Matilda Effect

Matilda Joslyn Gage was a nineteenth-century American human rights activist and pioneer in the fight for equal rights for women and Native Americans and the liberation of slaves. In 1993, historian Margaret Rossiter coined the term Matilda effect to describe the situation in which women scientists receive less credit for their scientific work than men with comparable achievements.

Ochoa became a naturalized US citizen in 1956 but maintained close ties with Spain. He continued to study enzymes and produced more than five hundred papers, most of them coauthored—a testament to the group nature of such detailed laboratory discoveries. He thoroughly enjoyed the work, stating, "In my life biochemistry has been my only and real hobby."[5] He was a friend of artist Salvador Dalí, who was commissioned to draw artwork for the cover of a 1976 book honoring Ochoa's scientific legacy. Dalí also wrote this homage to his friend:

> God does not play dice, wrote Albert Einstein long before the discovery of the DNA ladder, the steps of which are traced by angels in Jacob's dream that I had the night before painting this ladder for Severo Ochoa. These angels symbolize messengers of the genetic code, nucleotide molecules synthesized for the first time in Severo Ochoa's laboratory.
>
> Although I am not a scientist, I must confess that scientific events are the only ones that guide my imagination, at the same time that they illustrate the poetic intuitions of traditional philosophers to the point of achieving a dazzling beauty of certain mathematical structures, especially those of polytopes, and above all of those sublime moments

---

5   S. Ochoa, "The Pursuit of a Hobby," *Annual Review of Biochemistry* 49 (1980): 1–30.

of abstraction that, seen through an electron microscope, appear like viruses in a polyhedral form, confirming what Plato said: "God always makes geometry."[6]

Salvador Dalí, Polynucleotide Messengers (Tribute to Severo Ochoa) (1975). Drawing for lithography plate. (Reproduced with permission of Museo de Xixón, Asturias, Spain.)

Severo Ochoa returned to Spain at the end of his life and is regarded as a national hero there. He died in Madrid in 1993 at the age of eighty-eight.

---

6 S. Dalí, "My Homage to Severo Ochoa," in *Reflections on Biochemistry: In Honour of Severo Ochoa*, ed. A. Kornberg et al. (New York: Pergamon Press, 1976), 445.

One of Ochoa's students was Arthur Kornberg. Kornberg was born in New York City in 1918. He earned an MD at the University of Rochester then joined the coast guard as a ship's doctor, transferring within the public health service to the National Institutes of Health. There, he was assigned to study nutrition, in which capacity he helped to isolate vitamin B_9 (folate) and became interested in the role of enzymes in metabolic processes. He worked for a time in Severo Ochoa's lab as a postdoctoral researcher.

During the war years, Kornberg met and married Sylvy Ruth Levy, also an accomplished biochemist. According to their granddaughter, "Supposedly, on their first date, Arthur bragged to Sylvy that he had gotten a perfect score in chemistry on the New York State Regents Exam. He would later find out that Sylvy got perfect scores not only in chemistry, but also in geometry and algebra on the exams."[7]

By the time of Grunberg-Manago's RNA synthesis in Ochoa's lab, Kornberg was chairman of Washington University's Medical Microbiology Department in St. Louis. He built on Grunberg-Manago's results toward his project of attempting to make DNA in a test tube. While Grunberg-Manago was able to make RNA out of its amino acid building blocks, Kornberg found that he needed a template in order to make DNA. Some already-assembled DNA was required for the enzyme he discovered to extend the molecule and make more DNA. He won his half of the Nobel Prize for discovering the enzyme that did this DNA polymerase and was notified of the prize just after moving to Stanford to head the Department of Biochemistry there.

Sylvy Kornberg had worked closely with Arthur, making significant contributions to the discovery of the DNA enzyme that won him

---

7   M. Kornberg, "Telling Their Stories: A Tribute to the Life and Work of Sylvy Kornberg," SCOPE, the blog of Stanford Medicine, May 25, 2018, scopeblog. stanford.edu/2018/05/25/kornberg/.

the Nobel Prize.[8] Specifically, she identified and removed a contaminating enzyme that was interfering with their attempts at building DNA.[9] According to their son Thomas Kornberg, "The joke in the family—and it was just a joke—was that when the prize was announced, she said 'I was robbed!'"[10] In fact, Arthur Kornberg didn't mention Sylvy Kornberg in his Nobel lecture and cited only one of her many scientific papers.[11]

For the next twelve years Arthur Kornberg studied various ways to make a complete copy of DNA as it appears in nature. He and his teams published twenty-four papers on this topic, culminating in a 1967 publication announcing that they'd made a DNA virus.[12] Viruses are composed of short strands of either RNA or DNA, sometimes with a protein envelope. Using DNA polymerase, Kornberg's team made a very tiny DNA virus from its building-block proteins. It became headline news: a fully infectious virus had been made in a test tube. The big question became whether that meant humans had created a life form. Viruses are only sections of DNA or RNA and have no capacity to make more of themselves. In order to replicate, they need to penetrate the surface of a cell such as a bacterium or a human cell and inject genetic material (DNA or RNA) to take over that cell's machinery. Within minutes of doing so, they burst the cell to release hundreds of virus copies.

---

8 S. R. Kornberg, L. R. Lehman, M. J. Bessman, E. S. Simms, and A. Kornberg, "Enzymatic Cleavage of Deoxyguanosine Triphosphate to Deoxyguanosine and Tripolyphosphate," *Journal of Biological Chemistry* 223, no. 1 (1958): 159–62.

9 A. Kornberg, S. B. Zimmerman, S. R. Kornberg, and J. Josse, "Enzymatic Synthesis of Deoxyribonucleic Acid. Influence Of Bacteriophage T2 on the Synthetic Pathway in Host Cells," *Proceedings of the National Academy of Sciences of the United States of America* 45, no. 6 (1959): 772–85.

10 D. Kwon, "Sylvy Kornberg: Biography of a Biochemist," *Scientist*, June 13, 2017.

11 A. Kornberg, December 11, 1959, "The Biologic Synthesis of Deoxyribonucleic Acid," NobelPrize.org, nobelprize.org/prizes/medicine/1959/kornberg/lecture/.

12 M. Goulian, A. Kornberg, and R. Sinsheimer, "Enzymatic synthesis of DNA, XXIV. Synthesis of Infectious Phage *phiX174* DNA," *Proceedings of The National Academy of Sciences* 58, 2321–2328.

It was the beginning of the era of genetic engineering, but had they created life in a test tube? Kornberg hedged on the answer: "I tried to make it clear that in essence, there is no point at which the breath of life is infused in ascending from carbon atoms to nucleotides [building blocks of DNA] to DNA to viruses to cells to man."[13]

Among other contributions, Sylvy Kornberg discovered and characterized an enzyme, polyphosphate kinase (PPK), that is found in all living organisms,[14] which builds long chains of phosphate groups. Sylvy died in 1986, but Kornberg continued intensive research into PPK. Toward the end of his career, Kornberg worked with the ICOS Laboratory to identify sixty-two biochemicals that could inhibit the PPK enzyme.[15] Since PPK is found in all bacteria, Kornberg thought these could be developed into drugs that would treat a wide variety of infectious diseases. He focused on tuberculosis and was in negotiations to develop a new antibiotic for it.[16] At just that time (October 2006), ICOS was bought out by Eli Lilly. Lilly had been using ICOS to make their big-selling impotence drug, Cialis; once they were full owners, they dissolved ICOS, dedicating the lab's resources entirely to production of Cialis.[17] Research into PPK-related drugs virtually halted and has only been continued at a low level in other labs.

Kornberg had the immense satisfaction of seeing one of his sons also win the Nobel Prize. Roger Kornberg was awarded the 2006 prize in chemistry. His research built on the work of his father, Ochoa, and

---

13 A. Kornberg, *For the Love of Enzymes: The Odyssey of a Biochemist* (Cambridge, MA: Harvard University Press, 1981).

14 A. Kornberg, S. R. Kornberg, and E. S. Simms, "Metaphosphate Synthesis by an Enzyme from Escherichia coli," *Biochimica et Biophysica Acta* 20, no. 1 (1956): 215–27.

15 M. R. W. Brown and A. Kornberg, "The Long and Short of It: Polyphosphate, PPK and Bacterial survival," *Trends in Biochemical Sciences* 33, no. 6 (2008): 284–90.

16 A. Kornberg, Interview, December 2006, NobelPrize.org, nobelprize.org/prizes/medicine/1959/kornberg/interview/.

17 Eli Lilly and Company, "Lilly Announces Acquisition of ICOS Corporation" (press release), October 17, 2006, investor.lilly.com/news-releases/news-release-details/lilly-announces-acquisition-icos-corporation.

Grunberg-Manago. Working in the field of cells with nuclei, Roger Kornberg succeeded in mapping the process of exactly how information in DNA is transferred to RNA, and he determined the structure of the RNA polymerase enzyme.[18]

Arthur Kornberg died at the age of eighty-nine in 2007.

---

18 R. D. Kornberg, "The Molecular Basis of Eukaryotic Transcription," December 8, 2006, NobelPrize.org, nobelprize.org/prizes/chemistry/2006/kornberg/lecture/.

10

Self and Nonself

The 1960 prize went to Frank Burnet and Peter Medawar for deciphering how the immune system recognizes which tissues it should not attack. Their discoveries led to the understanding that in order to successfully receive an organ transplant, the recipient's immune system first must be suppressed so as not to reject the new organ. This prompted the development of powerful immune-suppressing drugs that made organ transplantation an everyday reality.

Frank Macfarlane Burnet was the second Australian to win the Nobel Prize in Medicine. He was born in Melbourne in 1899 and by his own account was an unconfident, awkward, and cowardly child.[1] He occupied himself with insect collecting and documenting the rich animal life found in the swamps near his home.

Burnet was a serious and ambitious student, but even though he received high marks in school, he was riddled with anxieties and self-doubt. After he studied the basics in medical school, Burnet's academic advisers guided him away from applying for a path training for surgery with live patients and urged him in the direction of laboratory research as a pathologist. That suited his introverted personality just fine.

---

1   C. Sexton, *The Seeds of Time: The Life of Sir Macfarlane Burnet* (Oxford: Oxford University Press, 1992).

Early in his research career, Burnet set his sights on winning the
Nobel Prize. He discovered that shingles in adults was caused by reactivation of the chicken pox virus after an initial childhood infection.[2]
He found the bacteria transmitted by sheep that was responsible for
fevers in slaughterhouse workers, which now carries his name, *Coxiella
burnetii*. The illness is called Q fever, and Burnet asserts that he named
it for Queensland, the Australian state where the fever was prevalent.
Other authorities say the Q came from the word *query*, since for so
long what caused the illness was unknown.[3] Q fever is still seen today
in animal handlers such as ranchers, farmers, and employees of livestock facilities. In 2014 there was an unusual cluster of Q fever in five
people in upstate New York. None of them were animal handlers, but
they had all recently been to Germany for experimental medical treatments with injections of fetal sheep stem cells. In most cases, people
with Q fever recover without treatment, but a minority of folks become
chronically ill with ongoing fever and fatigue, requiring several months
of treatment with the antibiotics doxycycline and chloroquine.[4]

Burnet did a great deal of experimental work on the viruses of polio
and influenza. His research brought him back and forth to England,
and on one such trip he used an old brown suitcase to transport vials
of potentially dangerous artificially modified virus particles. Twenty
years later, he wrote an article in the *Lancet* warning of the possibility
that manipulated viruses could easily escape into circulation and cause
a catastrophic epidemic in humans. His concern was that manipulated
viruses would be deadly because the human body would have had no
previous immunological experience with them, causing "the almost

2 F. M. Burnet and S. W. Williams, "Herpes Simplex: A New Point of View," *Medical Journal of Australia* 1 (1939): 637–42.

3 F. M. Burnet and M. Freeman, "Experimental Studies on the Virus of 'Q' Fever," *Medical Journal of Australia* 2 (1937): 299–305.

4 M. P. Robyn et al., "Q Fever Outbreak among Travelers to Germany who Received Live Cell Therapy—United States and Canada, 2014," *Morbidity and Mortality Weekly Report* 64, no. 38 (October 2, 2015): 1071–73.

unimaginable catastrophe of a 'virgin soil' epidemic . . . involving all the populous regions of the world."[5]

The theory of viral mutants escaping from labs is not outlandish. Today such infectious forms (viruses or bacteria) are called potential pandemic pathogens (PPPs), and there are clear historical precedents demonstrating that PPPs do "escape" from laboratories, even high-security facilities. The virus called H1N1 caused the Spanish flu epidemic of 1918 and then circulated in human populations for four decades; it disappeared in 1957 and remained absent for two decades but then reappeared in 1977. Gene sequences of the 1977 isolate compared to samples of the virus collected in the 1950s were almost identical, which tells us that the virus had not replicated or evolved in the interim.[6] Many labs are known to have held H1N1 in freezers between 1950 and 1977, and it is thought probable by many virologists that the virus escaped during routine culture processes.[7]

A second way that viruses can get out of the lab is within the cells in which they "incubate." One popular cell line used to grow viruses came from the kidney tissue of a cocker spaniel in 1958, the so-called MDCK cell line. MDCK cells are kept propagating for decades because they grow viruses very well and therefore make it convenient to study viruses. Uninvited viruses have been caught hitching a ride inside MDCK cells that are swapped from lab to lab.[8] Lab-created viruses have also contaminated cell lines used in HIV research.[9] Another type

5   F. M. Burnet, "Men or Molecules? A Hit at Molecular Biology," *Lancet* 287, no. 7427 (1966): 37–39.

6   A. P. Kendal, G. R. Noble, J. J. Skehel, and W. R. Dowdle, "Antigenic Similarity of Influenza A (H1N1) Viruses from Epidemics in 1977–1978 to 'Scandinavian' Strains Isolated in Epidemics of 1950–1951," *Virology* 89, no. 2 (1978): 632–36.

7   S. M. Zimmer and D. S. Burke, "Historical Perspective—Emergence of Influenza A (H1N1) Viruses," *New England Journal of Medicine* 361 (2009): 279–85.

8   A. Stang et al., "Unintended Spread of a Biosafety Level 2 Recombinant Retrovirus," *Retrovirology* 6 (2009): 68.

9   Y. Takeuchi, M. O. McClure, and M. Pizzato, "Identification of Gammaretroviruses Constitutively Released from Cell Lines Used for HIV Research," *Journal of Virology* 82, no. 24 (2008): 12585–88.

of cell line that is widely used in virus research was derived in the 1950s from mouse tissue; it was found to be contaminated with lab-created virus strains that can produce tumors in lab rats.[10]

Another route is the distribution of viruses introduced into livestock or humans through vaccination. For example, in 2010, DNA from lab-mutated pig virus was found in a rotavirus vaccine given to infants,[11] which was taken off the market for other reasons. In the case of the oral polio vaccine, there continues to be sporadic problems today. This vaccine is made with live but altered viruses, hopefully changed just enough to not cause illness but still similar enough to native poliovirus to provoke the desired immune response. Cases of vaccine-strain polio do occur rarely, especially in children who already have impaired immune systems.[12] Similarly, the altered measles virus that is an ingredient in the measles vaccine can cause infection, rarely resulting in brain inflammation and death.[13] Live-virus vaccines in livestock have been suspected to result in vaccine-strain viruses infecting human workers in piggeries and slaughterhouses.

Potential Pandemic Pathogens on the Loose

H5N1 was a newly discovered bird flu virus that killed two dozen people in Hong Kong in 1997. Between 2003 and 2020, there were 861 cases of H5N1 infection worldwide, causing 455 deaths.[14]

---

10 J. W. Hartley, L. H. Evans, K. Y. Green, et al., "Expression of Infectious Murine Leukemia Viruses by RAW264.7 Cells, a Potential Complication for Studies with a Widely Used Mouse Macrophage Cell Line," *Retrovirology* 5 (2008): 1.

11 S. M. Gilliland et al., "Investigation of Porcine Circovirus Contamination in Human Vaccines," *Biologicals* 40, no. 4 (2012): 270-7.

12 S. B. Troy et al., "Vaccine Poliovirus Shedding and Immune Response to Oral Polio Vaccine in HIV-Infected and -Uninfected Zimbabwean Infants," *Journal of Infectious Disease* 208, no. 4 (2013): 672-78.

13 A. Bitnun et al., "Measles Inclusion-Body Encephalitis Caused by the Vaccine Strain of Measles Virus," *Clinical Infectious Diseases* 29, no. 4 (1999): 855-61.

14 "Cumulative Number of Confirmed Human Cases for Avian Influenza A(H5N1) Reported to WHO, 2003-2020," World Health Organization, who.int/influenza/human_animal_interface/2020_01_20_tableH5N1.pdf?ua=1.

Although highly deadly, H5N1 doesn't pass easily from human to human. But scientists have genetically altered the virus to make it jump between species. Labs that work with the mutated virus are usually required to be biosafety level 3 (BSL3), which imposes relatively stringent safety precautions. In 2014, researchers from the Harvard and Yale schools of public health quantified the probability that this deadly super-flu virus with man-made mutations could be accidentally or deliberately released from a BSL3 lab: "From the conservative estimate of the rate of laboratory-associated infections of two per 1,000 laboratory-years, it follows that a moderate research program of ten laboratories at US BSL3 standards for a decade would run a nearly 20% risk of resulting in at least one laboratory-acquired infection, which, in turn, may initiate a chain of transmission. The probability that a laboratory-acquired influenza infection would lead to extensive spread has been estimated to be at least 10%. Simple branching process models suggest a probability of an outbreak arising from an accidental influenza infection in the range of 5% to 60%."[15] The researchers suggested an urgent reevaluation of whether experiments that artificially create animal pathways for the spread of deadly mutant viruses are really necessary.

In the lab, Burnet developed a weakened influenza strain and made it into a live-virus vaccine that could be given by nasal inhalation. It was first tested on medical student volunteers and Burnet's own ten-year old son before being given to a sample of army volunteers.[16] The vaccine appeared to be mostly harmless, so in 1942 it was used to vaccinate twenty thousand Australian troops. Although it had some

15 M. Lipsitch and A. P. Galvani, "Ethical Alternatives to Experiments with Novel Potential Pandemic Pathogens," *PLoS Medicine* 11, no. 5 (2014): e1001646.

16 F. M. Burnet, in "Written Address, 1942," as cited in C. Sexton, *The Seeds of Time: The Life of Sir Macfarlane Burnet* (Oxford: Oxford University Press, 1992).

effectiveness, a killed-virus version of an injectable flu vaccine was simultaneously being developed in America, and Australia soon lost interest in Burnet's live-virus version.

In a 1949 publication with coresearcher Frank Fenner, Burnet proposed a theory of how the immune system recognizes self versus nonself so that an organism will not attack its own cells but will appropriately attack similar cells from an outsider.[17] The theory proposed that if an organism is exposed to a particular type of foreign cell when it is still in a formative stage, then later in life it will not reject that kind of foreign cell and will tolerate it, as if it is part of itself. Burnet's experiments failed to prove the theory, which is where the young Peter Medawar entered the picture.

Peter Medawar was researching the phenomenon of how skin transplants were often rejected by recipients as their immune system sensed the foreign tissue and fought it off. He began a long-distance collaboration with Frank Burnet, which led to experiments wherein newborn mice of strain A received skin grafts from mice of strain B. Later in life, strain A mice did not reject skin grafts from strain B mice. This proved Burnet's theory and was the work that won them both the Nobel Prize.

It was already known that when immune cells meet a foreign substance (like a cell, virus, or bacterium), called an *antigen*, they are provoked to produce *antibodies* against the antigen. In 1957, Burnet was third author on a paper describing his theory of how the body recognizes self versus others.[18] He proposed that once a specific antigen is introduced, the body makes a clone of the corresponding specific antibody-making immune cells, expanding that cell's population by the millions. Burnet thought his "clonal selection theory" was worthy

---

17 F. M. Burnet and F. Fenner, *The Production of Antibodies*, 2nd ed. (New York: Macmillan, 1949).

18 I. R. Mackay, L. Larkin, and F. M. Burnet, "Failure of Autoimmune Antibody to React with Antigen Prepared from the Individual's Own Tissues," *Lancet* 273, no. 6986 (1957): 122–23.

of the Nobel Prize and regarded his earlier work as less important.[19] Nevertheless, it was the earlier research that was recognized in his nomination for the prize.

Burnet flourished after winning the Nobel, and like so many laureates he exploited his new stature to promote political and social philosophies that were far beyond his fields of expertise. He was invited to serve on numerous government and academic committees relating to the direction of scientific research and policies on public health and education in Australia. Burnet advocated for schools to scrap all literary studies as a waste of time. He advised that history should only be taught in the context of the lessons it illustrated about human behavior. He said, "Art and literature should be leisure time interests, and not something on which to waste the community's time and money in teaching."[20]

Burnet considered aggression and the desire for dominance to be innate in men and concluded that evil conduct was hardwired in genetics, dooming all attempts at social reform. Women, according to Burnet, were innately less aggressive, making them more likely to be innately good.[21]

Burnet earnestly believed in the righteousness of scientists as leaders, and he spoke on the necessity of prominent researchers to consider themselves world citizens, with their work impacting the fate of all mankind. These lofty views were incongruent with a darker streak, as revealed in Burnet's support of eugenics—a word he did not use. Instead, he referred to "genetically-based bonding," in which the best minds of society would breed in order to produce better human specimens. Burnet thought the "genetically stupid" were reproducing too rapidly, at the expense of the genetically intelligent. Burnet supposed

---

19 F. M. Burnet, *The Clonal Selection Theory of Acquired Immunity* (Nashville, TN: Vanderbilt University Press, 1959).

20 As quoted in C. Sexton, *The Seeds of Time: The Life of Sir Macfarlane Burnet* (Oxford: Oxford University Press, 1992).

21 F. M. Burnet, *Endurance of Life: The Implications of Genetics for Human Life* (Cambridge: Cambridge University Press, 1978).

that his own superior intelligence sprang from an inherited infrastructure of his nervous system.[22] He did not admit the inevitable liability of such a philosophy—Who decides on the "superior" qualities? One biography relates a social meeting wherein Burnet was the guest of Jawaharlal Nehru, the first prime minister of India. Burnet enthusiastically launched into unsolicited recommendations for population control, with no comprehension of the reason for the stony silence he received from Nehru. Burnet clearly considered the populations he advocated to be restrained not of his own class, country, or race.[23] Burnet supported selective infanticide, which would allow the euthanasia of babies born with incurable severe congenital defects. At the other end of life, he thought invasive medical devices and heroic procedures in the intensive care unit were a waste of resources on behalf of the aged. He advocated the death sentence for convicted rapists and murderers so as not to waste taxpayer money on long prison terms. Burnet also promoted the idea that scientists should work toward the progressive increase in the lethality of war. Like Alfred Nobel, Burnet reasoned that the thought of mutual annihilation would act as a powerful deterrent and thus bring about peace. On the other hand, he spoke out against Australian uranium mining and exports, which largely supplied nuclear warheads. He later modified his stance to that of supporting limited uranium exports, but only for peaceful uses in nuclear power plants.[24]

On the positive side, Burnet advised governments to require mandatory dire warnings on cigarette packages; these would finally become a reality a couple of decades later. With great foresight, he

---

22 F. M. Burnet, *Endurance of Life: The Implications of Genetics for Human Life* (Cambridge: Cambridge University Press, 1978).

23 F. M. Burnet, "Man's Responsibilities: Global Homeostasis" unpublished public address, May 1971, Box 27, Series 6, Records of Frank Macfarlane Burnet, University of Melbourne Archives, austehc.unimelb.edu.au/guides/burn/FMBS0006.htm.

24 F. M. Burnet, "The Implications of Global Homeostasis," *Impact of Science on Society* 22, no. 4 (1972): 305–14, October/December 72.

urged testing of emerging products and technologies for their impact on ecological systems.

———————

Peter Brian Medawar was born in 1915 in Brazil to a Lebanese father and British mother, and he later became a naturalized British citizen. Peter Medawar's personality was the extreme opposite of Frank Burnet. Medawar was outgoing, gregarious, and socially adept, and he had close friends all over the world.

In 1958, Medawar coauthored a paper describing the mouse skin graft experiments that demonstrated Burnet's theories.[25] Otherwise, the Nobel cowinners had profoundly differing opinions on several key issues. Medawar did not believe that a scientist's success gave him any special or privileged insight into solutions to social problems. In a BBC interview, Medawar said, "Scientists are sometimes inclined to feel that if only they could have possession of great political authority and exercise the scientific method then all problems would dissolve before their acute critical gaze. It does not work like that, it just is not true."[26] Medawar called out self-important scientists, such as James Watson, cowinner of the 1962 Nobel, whom he described as "having a somewhat messianic conception of his role in the great revolution of molecular genetics."[27] Yet he did not deny Watson's insightful scientific contributions.

Medawar thought that fears of the overwhelm of humans by new and strange viruses reflected a misunderstanding of actual physiological facts. He explained, "There is after all always a first time we are confronted with any disease-causing organism, but we do not necessarily succumb to it."[28] Medawar also had something to say about

———————

25 P. B. Medawar and P. B. Woodruff, "The Induction of Tolerance by Skin Homografts on Newborn Rats," *Immunology* 1, no. 1 (1958): 27–35.

26 As quoted in P. Medawar, *The Threat and the Glory: Reflections on Science and Scientists* (Oxford: Oxford University Press, 1991).

27 P. Medawar, *The Threat and the Glory* (Oxford: Oxford University Press, 1991).

28 P. Medawar, *The Threat And the Glory* (Oxford Oxford University Press, 1991).

social inbreeding for the supposed betterment of the human race. He explained that the concept of eugenics was scientifically unfounded and in fact rendered impossible by the sheer variety of genetic variants. Natural selection does not work toward the fixation of particular characteristics in a specific pedigree. Instead, he explained, it is the whole population of Earth that evolves.[29] Medawar also held a very different opinion of heroic medical interventions, having sustained a number of strokes that had rendered him temporarily dependent on life-saving measures.[30]

Medawar observed that scientists are a varied bunch who cannot be easily classified, with temperaments ranging from detective-like to mystical or philosophical, "with a few odd crooks."[31] Neither Burnet nor Medawar can be described by any single attribute, but the prize-winning pair is an example of the wide diversity in Nobel laureates.

Frank Burnet died in 1985 at the age of eighty-five. Peter Medawar died in 1987 at the age of seventy-two.

---

29 P. Medawar, "The Future of Man," in *The Threat and the Glory* (Oxford: Oxford University Press, 1991).

30 P. Medawar, "Son of Stroke," in *The Threat and the Glory* (Oxford: Oxford University Press, 1991).

31 P. Medawar, "An Essay on Scians [*sic*]," in *The Limits of Science* (New York: Harper and Row, 1984).

11

Postal Worker Wins the Nobel Prize

Georg von Békésy won the 1961 Nobel Prize in Medicine for discovering how sound stimulates the inner ear. He was born in 1899 in Budapest, Hungary, and first studied chemistry at the University of Bern in Switzerland then physics at the University of Budapest. By the time he had his doctorate in physics in the early 1920s, the country was impoverished in the aftermath of WWI and there were no well-equipped and well-funded university physics laboratories in which to work. He took a job with the national post office, which doubled as home to the country's national telephone operations, where he was charged with improving the quality and speed of electronic communications systems. This led him to study how humans perceive sound.

The first problem he tackled was how to get a good look at the inner ear of humans. Most ear research had been done on animals, because a human cadaver tends to rapidly dry out, hardening the delicate soft structures of the inner ear. Georg von Békésy solved this by obtaining only the freshest corpses. He also devised a way to rapidly dissect the ear while keeping the tissues moist. This involved a low-power microscope and a bone saw that operated while the decapitated head was held in a water bath—a setup so noisy and objectionable that it drew complaints from fellow postal workers and telephone researchers.

His dissection methods left intact the tiny inner part of the ear that he needed to study. The hearing part of the ear is rolled up into a spiral like a snail shell, so it is called the cochlea, from the Greek for snail. A thin membrane runs in the middle of the cochlea along the length of the spiraled chamber, which is otherwise filled with fluid. While the inner-ear structures could be seen well enough to describe the anatomy, they were so small that it was still difficult to study how the ear dynamically transmitted sound sensations. A novel solution was needed, and, as von Békésy himself related, he never liked to work too hard.[1] His physics background gave him a very practical approach to problems, and he ended up making a large tabletop model of an unwound cochlea. He discovered that sound causes the membrane to undulate in a wavelike fashion, like a rope being snapped. High frequency sound sets up a series of waves that peaks on the membrane portion nearest the base of the cochlea, and low frequency sound causes waves that peak on the membrane closer to the apex, or tip, of the cochlea. In the human ear, this wave motion stimulates the nerve cells lining the cochlea, which transmit the energy into electrical signals that go to the brain to be recognized as sound.

However, von Békésy could not rig up model nerves and a brain, so he needed to substitute another sensory organ to receive the stimulus transmitted through the model ear. In order to do that, he did what he called "trying to do science in an unscientific way."[2] His solution to the problem was to position his forearm alongside the tabletop ear machine. It worked—he could feel different qualities of sensation on

1   G. von Békésy, "Concerning the Pleasures of Observing, and the Mechanics of the Inner Ear," in *Nobel Lectures Including Presentation Speeches and Laureates' Biographies: Physiology of Medicine 1942–1962* (Amsterdam: Elsevier, 1964), 722–46.

2   G. von Békésy, "Concerning the Pleasures of Observing, and the Mechanics of the Inner Ear," in *Nobel Lectures Including Presentation Speeches and Laureates' Biographies: Physiology of Medicine 1942–1962* (Amsterdam: Elsevier, 1964), 722–46.

the skin of his arm depending upon the frequency of sound that came through.[3] His own description says it best:

> The final version of the model consists of a plastic tube filled with water, and a membrane 30cm in length; when it is stimulated with a vibration it shows travelling waves of the same type as those seen in the normal human ear . . . I simply placed my arm against the model. To my surprise, although the travelling waves ran along the whole length of the membrane with almost the same amplitude, and only a quite flat maximum at one spot, the sensations along my arm were completely different. I had the impression that only a section of the membrane, 2 to 3 cm long, was vibrating."[4]

In other words, the sense organ (in this case his arm) only perceived the high point, where the wave was at its peak; the rest of the smaller waves were not perceived.

The Nobel lecture von Békésy gave is unusual in that he included in his presentation several images from his extensive and eclectic art collection interspersed with the usual scientific graphs and tables. Among other objects, he presented a circa-1400 BC Egyptian baboon sculpture and a figurine of a running fox from eighth-century-BC Greece to illustrate the timeless nature of communication in art. He described spending hours and hours gazing at a single piece of art as he pondered how to imagine something new—specifically, how is it possible to produce new discoveries in science when our imagination is so limited? The answer he found was not from chemistry or physics textbooks but in the works of da Vinci: "Da Vinci did not try to

---

3   G. von Békésy, "Simplified Model to Demonstrate the Energy Flow and Formation of Traveling Waves Similar to Those Found in the Cochlea," *Proceedings of the National Academy of Sciences of the United States of America* 42, no. 12 (1956): 930–44, doi.org/10.1073/pnas.42.12.930.

4   G. von Békésy, "Concerning the Pleasures of Observing, and the Mechanics of the Inner Ear," in *Nobel Lectures Including Presentation Speeches and Laureates' Biographies: Physiology of Medicine 1942–1962* (Amsterdam: Elsevier, 1964), 722–46.

outdo Nature with his fantasy, but, quite the opposite, he tried to learn from Nature."[5]

So, von Békésy decided that he needed to observe nature, leading to some of his most fascinating discoveries, well after his prizewinning research. The successful substitute of his arm for a brain in order to elucidate the sensory mechanism of the ear led to a series of experiments over the next few decades that examined the similarities among how humans perceive differing sensations, specifically vision, smell, taste, and touch. He researched for a year at the Karolinska Institute, which helped him to become known to those who would be making Nobel Prize decisions in the future. Then he was on the faculty of Harvard for nearly twenty years.

One of his areas of intense research was on Mach bands, the name given to the regions of decreased sensation that surround an area of increased sensation. For example, when a light hits the eye there is relatively increased sensitivity at a point, while the body responds by inhibiting sensitivity in the regions immediately surrounding it. This was initially described by Ernst Mach in 1865.[6]

To generalize the concept, von Békésy showed that Mach bands are exhibited in other sensory systems, too. In a simple demonstration of Mach bands in hearing, blaring loudspeakers were set up in a row. A person walking down the row of speakers did not hear sound from all the speakers, as might be expected. Instead, they heard the speaker nearest them, while not perceiving sound from the other speakers. They got a sense of someone alongside them carrying a speaker as they walked down the row together. The sensitivity to one stimulus (the nearest speaker) causes a complementary inhibition of sensitivity to

---

5   G. von Békésy, "Concerning the Pleasures of Observing, and the Mechanics of the Inner Ear," in *Nobel Lectures Including Presentation Speeches and Laureates' Biographies: Physiology of Medicine 1942–1962* (Amsterdam: Elsevier, 1964), 722–46.

6   R. B. Lotto et al., "Mach Bands as Empirically Derived Associations," *Proceedings of the National Academy of Sciences* 96, no. 9 (1999): 5245–50.

more distant speakers.[7] This explains why in his cochlear model experiments von Békésy felt vibration in his arm only in one region (at the point of peak intensity), even though the whole membrane was seen to be vibrating.

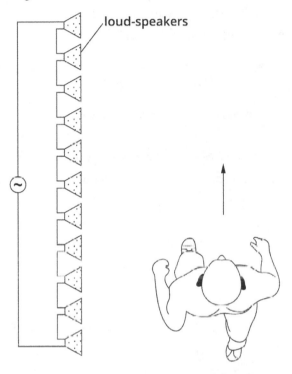

loud-speakers

A row of loudspeakers supplied by the same electrical source does not produce a longitudinal sound image. The impression is that only one of the loudspeakers is activated. The others seem to be suppressed."[8]

His analogous experiment for the sensation of taste involved placing a drop of dilute acid solution on the tongue. This caused inhibition of sensing any taste at all in the immediately adjacent areas of the

---

7   G. von Békésy, "Mach Band Type Lateral Inhibition in Different Sense Organs," *Journal of General Physiology* 50, no. 3 (1967): 519–32.

8   G. von Békésy, "Mach Band Type Lateral Inhibition in Different Sense Organs," *Journal of General Physiology* 50, no. 3 (1967): 519–32.

tongue. He demonstrated the same phenomenon for sensations in the skin, having subjects rest their arms lightly on a section of stretching rubber tubing, causing a shearing force to be perceived. There was no shearing sensation at all in the areas immediately adjacent to the main stimulus. Thus, von Békésy concluded that for every nerve stimulus, there is corresponding nerve inhibition.[9]

The Amazing Ear

Sound is caused by a vibrating object and carried to your ears by the cyclic movement of the air. Its frequency can be measured in Hertz, a unit of measurement of repeating cycles, named after Heinrich Hertz and abbreviated Hz. The frequency of anything that is cyclic can be measured in Hertz.

High-frequency sound has cycles that are repeated rapidly, with shorter distances between wave peaks that move through the air, and it is perceived as high pitched. Low-frequency sound has cycles that are repeated more slowly, with longer wavelengths moving through the air, and is perceived as low pitched. The exact spacing between sound waves is highly dependent upon the air temperature: at 0° centigrade, 20 Hz has a cycle that is 16.5 meters long.

Perfect human hearing has the potential to perceive sound between 20 Hz and 20,000 Hz. Twenty Hz is about an octave below the lowest note on a bass guitar. One instrument that can achieve that is the low base pedals of an organ. It is difficult to hear because the wavelength is so long that even minor air disturbances can distort it. You may feel a throb in your body more than hear it.[10] Sounds higher and lower than this range do exist, humans just can't hear them. Ferrets and elephants can hear even lower ranges,

---

9 G. von Békésy, "Mach Band Type Lateral Inhibition in Different Sense Organs," *Journal of General Physiology* 50, no. 3 (1967): 519–32.

10 A demonstration of signal frequency with sound can be seen at "Human Audio Spectrum," Science and Technology, October 2, 2012, clip-share.net/video/qNf9nz vnd1k/20hz-to-20khz-human-audio-spectrum.html.

while higher ranges can be heard by dogs, cats, mice, sheep, ferrets, and many other creatures. The highest known hearing ranges
are in porpoises, who can hear 150,000 Hz, followed by beluga at
123,000 Hz and bats at 110,000 Hz.

Another area of von Békésy's research was in adaptation, whereby
an initial stimulus is perceived at a relatively low intensity, but when
the stimulus is repeated, it requires higher intensities to be perceived.
He demonstrated that adaptation also occurs for taste, again using a
dilute acid solution dripping on the tongue. At low concentrations,
the observer could taste it, but when the acid concentration increased
slightly, it took longer for the observer to taste it. Repeated over
and over, it took increasingly longer to taste the more concentrated
solutions.[11]

Von Békésy wondered if the well-known ability of the human lis-
tener to detect the direction of the source of a sound also held true
for the source of an odor. A person with two ears can perceive that
clicks come from different directions down to 0.1 millisecond, that
is, the clicks being only one ten-thousandth of a second apart. Von
Békésy showed that the nostrils can perceive the direction of sources
of odors also down to 0.1 millisecond, and this held true for a range
of smells from benzol, with a sweet gasoline-like chemical odor, to the
flowery hints of lavender.[12] Then it was shown that the specific tongue
locations one is tasting with can be perceived with accuracy when the
stimuli of flavored water are dropped on different regions of the tongue
as close as 0.1 millisecond apart. Von Békésy demonstrated that, in
the same manner, vision direction can be differentiated (as long as

11 G. von Békésy, "The Effect of Adaptation of the Taste Threshold Observed with a
    Semiautomatic Gustometer," *Journal of General Physiology* 48, no. 3 (1965): 481–88,
    doi.org/10.1085/Jgp.48.3.481.

12 G. von Békésy, "Olfactory Analogue to Directional Hearing," *Journal of Applied
    Physiology*, May 1, 1964.

the light stimulus is sweeping rather than pinpoint).[13] Von Békésy predicted that sensory systems, such as vision, hearing, smelling, and touch, which are treated in textbooks as unrelated entities, would be addressed in the future under one combined discipline—this has not yet come to pass.

In the modern era, he is described as a biophysicist, one who applies the principles of physics to the problems of living systems. Today's biophysicists are involved with everything from studying how bodies react to weightlessness in space to how molecules interact with one another.

Von Békésy was a loner and had no immediate family, his objects of art being his most consistent companions. He had objects from Persia, India, Egypt, Nigeria, Japan, Mexico, Peru, Cambodia, Thailand, China, and Tibet, ranging from the sixth century BC into the twentieth century AD. A colleague from Harvard described the role of art for von Békésy:

> His closest friends, from which he drew both solace and inspiration, were the art objects he had collected over the years. These filled his laboratory, secreted here and there in drawers and filing cabinets where one might ordinarily expect to find only tools, supplies, and records of data. But always at least one of these treasures was out on display on his work bench or desk where he might spend hours examining it and reflecting on its beauty.[14]

Von Békésy faced mandatory retirement from Harvard at the age of sixty-seven, but he continued active research at a sensory-sciences lab built especially for him at University of Hawaii. He died in 1972.

---

13 G. von Békésy, "The Smallest Time Difference the Eyes Can Detect with Sweeping Stimulation," *Proceedings of the National Academy of Sciences* 64, no. 1 (1969): 142–47, doi.org/10.1073/Pnas.64.1.142.

14 F. Ratliff, "Georg von Békésy, 1899–1972, A Biographical Memoir," (Washington, DC: National Academy of Sciences ,1976).

12

The Wonder Boys

Maurice Wilkins, Francis Crick, and James Watson shared the Nobel Prize in 1962 for their discoveries concerning the substance that carries our genetic information.

By the time of their prizewining research in the early 1950s, DNA was already known to be the stuff that somehow transmits genetic information. It had been determined to be a very long chain (or chains) composed of ordinary molecules—sugars, phosphates, and four kinds of nitrogen-containing compounds, which were called bases. There was nothing special about the ingredients: sugars are simply made of carbon and water; phosphates are found everywhere in the body, most abundantly in the teeth and bones, and the nitrogen in the bases was part of ring-shaped molecules made mostly of carbon.

Scientists around the world competed to be the first to figure out the exact structure of DNA, a structure that could carry out the function of providing the instructions for making all the various substances to build, maintain, run, and repair a body. Although the individual ingredients were known, it was not understood how they fit together in three dimensions. It was like having flour, eggs, and yeast with no knowledge of how to combine them in recipes for endless variations of baked goods.

Maurice Wilkins was born in New Zealand to Irish parents, and he was raised and educated in England. He received a degree in physics just as WWII started, so he was recruited into helping improve radar in England, then he worked on the British atomic bomb project, separating radioactive forms of uranium. This latter work was continued when he joined the Manhattan Project for development of the nuclear bomb in the US.

After the war, Wilkins researched many different approaches to determining the molecular structure of biological substances. While at King's College in London, he looked at DNA by making it into a crystal form and then striking the crystal with narrow beams of X-rays. This method created pictures that gave hints to the 3-D arrangement of DNA's chemical components. X-ray crystallography had been developed in the 1930s and was also being used by others at King's College, including the chemist Rosalind Franklin and a graduate student she was supervising, Raymond Gosling.

Gosling produced a series of X-ray photographs of DNA and showed his fifty-first photo to Wilkins, who recognized that it suggested that the structure of DNA was a helix, or a twisted strand. A helical shape for DNA had been proposed a few years before, but the photo provided concrete evidence. Wilkins showed the photo to his friend Francis Crick at Cambridge, who in turn brought it to his colleague James Watson. Crick and Watson never handled DNA in the lab, but with the information provided by "Photo 51," they used slide rules and mathematical formulas to work out its structure: two stands woven in a twisting form—the double helix—with backbones of sugars and phosphates, the inside being the bases arranged in pairs.

The April 25, 1953, issue of the prestigious scientific journal *Nature* carried three back-to-back articles on the topic. Crick and Watson wrote a one-page report laying out the structure of DNA and suggesting how that specific structure allowed for the coding of genes.[1]

---

1   J. D. Watson and F. H. Crick, "A Structure for Deoxyribose Nucleic Acid," *Nature* 171 (1953): 737–38.

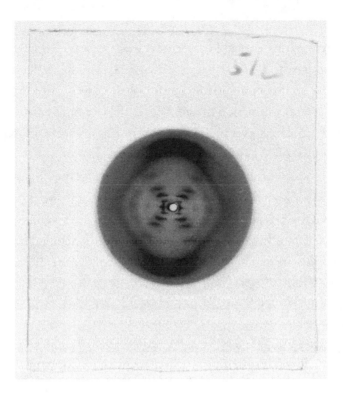

The next article was by Wilkins and colleagues on X-ray pictures of a related molecule, RNA,[2] which copies sections of the code from DNA then transports it to the cell machinery to make proteins. The final arti cle was by Franklin and Gosling, describing the X-ray Photo 51.[3] The elucidation of the basic structure of DNA was a turning point in biol ogy, eventually making possible the molecular mapping of every gene.

Much has been written about the contribution of Rosalind Franklin to the prizewinning discoveries, with some concluding she "was airbrushed out of the greatest scientific discovery of the twentieth century."[4] Today's Nobel Prize organization website credits Franklin

2   M. H. F. Wilkins, A. R. Stokes, and H. R. Wilson, "Molecular Structure of Deoxypentose Nucleic Acids," *Nature* 171 (1953): 738–40.

3   R. E. Franklin and R. G. Gosling, "Molecular Configuration in Sodium Thymonucleate," *Nature* 171 (1953): 740–41.

4   Summary copy for *Rosalind Franklin: The Dark Lady of DNA*, Harpercollins.com, www.harpercollins.com/products/rosalind-franklin-brenda-maddox.

with this blurb on their page about Wilkins: "Maurice Wilkins and Rosalind Franklin worked to determine the structure of the DNA molecule in the early 1950s at King's College in London. While they did not succeed in mapping the structure, their results—not least of all Franklin's X-ray diffraction images—were important in Francis Crick's and James Watson's eventual unlocking of the mystery: a long spiral with twin threads."[5] Gosling is never mentioned.

But it was Gosling who produced the important Photo 51, and although Franklin had seen it, she did not deduce the structure of DNA from it—in fact, she put the photo aside to study a different crystal form of DNA. In her 1953 *Nature* article, she concluded that Photo 51 supported the model proposed by Crick and Watson. The most that can be said is that Rosalind Franklin, as likely as many others working on the problem, would have eventually solved the structure of DNA if Watson and Crick had not seen Gosling's photo and come up with the model first. In other words, Watson and Crick beat everyone else to it. Franklin was never under consideration for the 1962 Nobel Prize for an entirely different reason: she'd succumbed to ovarian cancer in 1958.

At about the same time as his prizewinning research, Maurice Wilkins was under investigation by the British security service MI5 on suspicion of giving atomic secrets to the Soviets. Documents declassified in 2010 revealed that the US Federal Bureau of Investigation gave information to MI5 that someone who was on the international team of scientists working in the US with the Manhattan Project in the 1940s, "possibly an Australian or New Zealander," had passed secret information to the American Communist Party. Over the course of a three-year investigation, Wilkins's mail was intercepted and his activities were tracked. Wilkins had been born in New Zealand, later becoming a naturalized British citizen. He came to the attention of the security service because he was said to have defended the actions of British scientist Alan Nunn May, a fellow physicist at King's College

---

5   "Maurice Wilkins: Facts," NobelPrize.org, nobelprize.org/prizes/medicine/1962/wilkins/facts/.

who had confessed to passing atomic research secrets to the Soviets during World War II. Nunn May was released after six and a half years of a ten-year sentence to hard labor. To his dying day, Nunn May refuted the charge of treason, defending his actions as an earnest anti-German effort. According to the declassified files, Wilkins was "vehement in the defence of (Alan) Nunn May and the right of physicists to pass on their knowledge as they saw fit."[6]

Wilkins also knew fellow physicist Klaus Fuchs. Fuchs was German born and a former member of the German Communist Party who fled to England in 1933 when communists were being hunted in the early days of Nazi Germany. At the outbreak of WWII, Fuchs was briefly confined with other aliens by British authorities then released to continue his studies at the University of Birmingham. There he worked on the British atomic bomb project and came into contact with Wilkins. In 1943, he joined the Manhattan Project at Columbia University in New York, where he worked on uranium enrichment before he eventually landed a position at Los Alamos working on the plutonium bomb. In 1950, Fuchs confessed to passing atomic secrets to the Soviets during and after the war. Aside from these affiliations, there was no evidence on Wilkins, and the investigation was closed.

Later in life, Wilkins became a strong antinuclear campaigner. He frequently attended Pugwash conferences, the international meetings of scholars and public figures who seek to reduce the danger from armed conflict, with special attention on nuclear threats. Wilkins supported the British organizations Scientists Against Nuclear Arms and the Campaign for Nuclear Disarmament. Wilkins was founding president of the British Society for Social Responsibility in Science (cowinner Francis Crick was also a member), and he served as president of Food and Disarmament International from 1984 until his death.

---

6   As quoted in R. Stevens, "Kiwi Nobel Laureate Maurice Wilkins Investigated by MI5 in the 1950s for Spying," Stuff, December 31, 2015, stuff.co.nz/science/74734722/ kiwi-nobel-laureate-maurice-wilkins-investigated-by-mi5-in-the-1950s-for-spying.

Scientists as Social Activists

The Russell-Einstein Manifesto was a statement drafted in 1955 by Albert Einstein (1921 winner of the Nobel Prize in physics) and Bertrand Russell (1950 winner of the literature prize). Signed by Einstein and Russell along with nine other Nobel laureates and Polish physicist Leopold Infeld, it called for a nonpolitical meeting of world scientists to discuss the dangers of weapons of mass destruction. A few days after the release of the manifesto, the US-based businessman Cyrus Eaton offered to sponsor such a conference in Pugwash, Nova Scotia, his birthplace. The Pugwash Conferences on Science and World Affairs have been held regularly since 1957.

Scientists Against Nuclear Arms (SANA) was formed in England in 1981 by physicist Mike Pentz and neuroscientist Steven Rose. SANA was one of the forerunner organizations of Scientists for Global Responsibility (SGR). SGR came about in 1992 as a merger of SANA with Electronics and Computing for Peace and Psychologists for Peace, and was later joined by Architects and Engineers for Social Responsibility. Today SGR has a broader scope of social concerns, including advocating careers that emphasize peace, social justice, and environmental sustainability, and reducing military and commercial influences on science and technology.

The Campaign for Nuclear Disarmament was launched in 1958 as a platform for peaceful protests advocating the moral duty of Great Britain to disarm, unilaterally if necessary, with a broader goal of a global nuclear weapons ban.

The British Society for Social Responsibility in Science was formed in 1968 to oppose the development of chemical and biological weapons. It disbanded in the early 1990s.

Food and Disarmament International was founded in 1981 with the aims of eradicating starvation in the world and ensuring the basic rights to be free from hunger and armaments.

Francis Crick was born 1916 in England. He joined the Cavendish Laboratory of Cambridge University to work on his PhD in theoretical physics in the area of X-ray crystallography, a study that rarely required any lab work and was mostly accomplished with mathematical formulas. There he met the visiting American James Watson. Crick and Watson were keenly aware of the international race to discern the structure of DNA, and they followed the scientific publications on the topic with intense interest. Crick in particular was quick to spot the flaws in the math and models proposed by others. The upset caused by Crick's harsh assessments provoked the head of the lab to ban Crick and Watson from working on DNA for a time. Nevertheless, they continued to work on the problem of DNA structure. In 1952, they came up with a three-stranded model that didn't work.

Photo 51 gave them definitive direction. Crick and Watson used the clues to DNA size and structure gleaned from Photo 51 to refine their model. They came up with a precise description of a ladder-like double helix, the backbones running in opposite directions and composed of repeating units of sugars and phosphates, the "steps" being paired bases. The precise arrangement was determined by complex measurements of the angles formed by the various chemical bonds. These were fortified by detailed mathematics based on interpretations that Crick had developed in his earlier PhD thesis on horse blood. The new model not only complied with physical chemistry laws of how molecules fit together, it provided an explanation of how the base pairs could be combined in endless combinations to code for millions of proteins. They even built a model on their desktop to prove out their calculations.

An ordinary mortal can stare at Photo 51 for years and perceive only a blurry pattern of dashes roughly in the shape of an X. Apparently the image is only something that molecular physicists can get excited about, as it allows them to measure the discreet angles at which X-rays had been deflected by the DNA molecule. To others, it bears no relationship to the final picture of DNA illustrated in the 1953 *Nature* article as drawn by Crick's artist wife, Odile, which has been called "the most famous [scientific] drawing of the 20th century."[7]

Crick enjoyed his celebrity status as a Nobel laureate, flaunting his accomplishment by adorning the outside of his house in Cambridge with a yellow-painted brass spiral affixed above the front door. The Cricks hosted parties regularly, and according to one biographer, "A typical party . . . organised on the slightest pretext, would fill all four floors of the Golden Helix with friends, music, punch bowls . . . and the scent of the odd joint in the air."[8] Although Crick admired the writings of psychedelic-drug-advocate Aldous Huxley, rumors of his use of LSD cannot be confirmed.

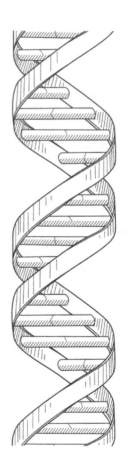

Upon figuring out the structure of DNA, Crick is said to have walked into the Eagle, a pub in Cambridge, and proclaimed he and James Watson had "found the secret of life." He soon realized they'd only opened the door to additional questions, such as, where did DNA come from. Crick observed that there are complex systems common to all cellular life, including a common genetic code that varies little between species and a common mechanism to translate it into proteins, which in turn help to reproduce the DNA code. These facts suggested to Crick

---

7   T. Sejnowski, as quoted in A. Bernstein, "Odile Crick, at 86; Artist Sketched Nudes and First Structure of DNA for Husband," *Washington Post*, July 22, 2007.

8   M. Ridley, *Francis Crick: Discoverer of the Genetic Code* (New York: HarperCollins, 2006).

that life on Earth originated from a small "bottleneck" population. He also thought it was impossible for DNA, the code mechanism, to magically evolve at the exact same time as RNA, the code copier, messenger, and translator.

Crick shared this interest with a colleague, the British chemist Leslie Orgel. The two were attending a meeting in Soviet Armenia in 1971 on the topic of communicating with aliens, when they struck on the theory of "directed panspermia." The word *panspermia* was coined by physicist Svante Arrhenius in 1907 to describe the theory that microorganisms drifted to Earth through space and seeded all life. Directed panspermia modified the basic concept to suggest aliens had deliberately sent the microbes to Earth. In 1973, Crick and Orgel published their theory in a scientific journal.[9] Orgel subsequently downplayed this line of reasoning, instead devoting his career to speculation about how the building blocks of life spontaneously emerged from early-earth chemicals such as frozen cyanide. However, Crick remained interested, and in 1981, he wrote what he called "a scientific book for the layperson" describing the scientific support for directed panspermia.[10]

Later in life, Crick worked at the Salk Institute in California, where he mused on the theoretical aspects of brain biology, although he did not involve himself in the conduct of any significant experiments. In his biography, Crick admitted that scientific facts about the brain were clear as mud: "It is very easy to convince someone that however he may think his brain works, it certainly doesn't work like that." In studying how the brain processes visual information, he said, "Another thing I discovered was that although much is known about the behavior of neurons in many parts of the visual system (at least in monkeys), nobody really has any clear idea how we actually see anything at all." His research on the anatomical location and biochemical basis of consciousness likewise went nowhere. "I wish I could say that my own

9   F. H. C. Crick and L. E. Orgel, "Directed Panspermia," *Icarus* 19, no. 3 (1973): 341–46.
10 F. Crick, *Life Itself: Its Origin and Nature* (New York: Simon and Schuster, 1981).

efforts amounted to much."[11] Crick died in California in 2004 at the age of eighty-eight.

———

James Watson was born in Chicago in 1928. At the age of fifteen, he was awarded a scholarship to the University of Chicago; he earned a PhD from Indiana University by the time he was twenty-two. His early work was on the subject of viruses that attack and infect bacteria. He next experimented with the X-ray crystallography of DNA at Cavendish Laboratory. In 1951, two years before he glimpsed Photo 51, Watson attended a scientific presentation by Rosalind Franklin in which she shared much of her data on DNA crystallography. Despite the information at their fingertips, Franklin didn't grasp the DNA structure, and neither did Watson—until, that is, in 1953 when he saw the picture and built on the work by Wilkins and collaborated with Crick.

A few years after winning the Nobel Prize, Watson joined the Harvard faculty. For the next twenty years there, he championed the focus of the department on molecular biology. He wrote textbooks, he wrote about himself, and he wrote about science, but he did not make any major scientific contributions.

Watson credits his early years at the University of Chicago with instilling in him "an ethical compulsion not to suffer fools."[12] By most accounts, the young Watson was a brilliant scientist, but also aggressively ambitious, hugely self-important, racist, and sexist to boot. For example, biologist E. O. Wilson of Harvard University described Watson as "radiating contempt" for the rest of the staff when he arrived there. "Having risen to fame at an early age, [Watson] became the Caligula of biology. He was given license to say anything that came into his mind and expected to be taken seriously. And unfortunately he

———

11 F. Crick, *What Mad Pursuit: A Personal View of Scientific Discovery* (New York: Basic Books, 1988).

12 J. D. Watson, *Avoid Boring People: Lessons from a Life in Science* (Oxford: Oxford University Press, 2007).

did so, with casual and brutal offhandedness."[13] The two later became congenial colleagues.

These attributes were known by friends and coworkers but became obvious to people who'd never met the man when his memoir, *The Double Helix*, became a bestseller in 1968. In the book, Watson makes deprecating remarks about various colleagues, including Rosalind Franklin. He wrote, "By choice she did not emphasize her feminine qualities. Though her features were strong, she was not unattractive and might have been quite stunning had she taken even a mild interest in clothes. This she did not. There was never lipstick to contrast with her straight black hair, while at the age of thirty-one her dresses showed all the imagination of English blue-stocking adolescents. So it was quite easy to imagine her the product of an unsatisfied mother who unduly stressed the desirability of professional careers that could save bright girls from marriages to dull men."[14] The book was initially supposed to be published by Harvard University, but they dropped the project, in part due to objections by Crick and Wilkins, who found it to be gossipy and not really about the science.

Nor was Watson shy about sharing his ideologies on improving the gene pool. In 1997 he suggested it would be acceptable to terminate a fetus if genetic testing (in the future) could identify that the baby would turn out to be homosexual. His later attempt to clarify the matter only dug him in deeper: "During an interview, I was asked about homosexuality and I related a story about a woman who felt her life had been ruined because her son was a homosexual and she would never have grandchildren. I simply said that women in that situation should have a choice over whether or not to abort. I didn't say that fetuses found to have a gay gene should be aborted." Further, Watson felt that a woman should have the option to abort if the fetus carries genes predicting

13 E. O. Wilson, *Naturalist* (Washington, DC: Island Press, 1994).
14 J. Watson, *The Double Helix: A Personal Account of the Discovery of the Structure of DNA* (New York: Atheneum, 1968).

dyslexia or a lack of musical ability (no such genes have been found) or that the child would be too short to play basketball.[15]

In 2000, Watson was guest lecturer at the University of California, Berkley, when he drew a link between sexual activity and skin color. An attendee at his lecture said Watson "showed slides of women in bikinis and contrasted them to veiled Muslim women, to suggest that controlling exposure to sun may suppress sexual desire and vice versa." Watson said, "That is why you have Latin lovers. You've never heard of an English lover." At the same presentation, he said that thin people are more ambitious and therefore make better workers, the flip side being that "whenever you interview fat people, you feel bad, because you know you're not going to hire them."[16] In an interview aired in 2003, Watson suggested that stupidity was a disease that could be eliminated in the future with gene editing, wherein suspect genes would be snipped out (no such genes have been identified). "The lower 10 per cent who really have difficulty, even in elementary school, what's the cause of it? A lot of people would like to say, 'Well, poverty, things like that.' It probably isn't. So I'd like to get rid of that, to help the lower 10 per cent."[17]

Meanwhile, Watson was affiliated with the National Institutes of Health (NIH) for four years, where he directed the Human Genome Project from 1990 to 1992. The genome is the entire set of genetic instructions found in a cell, consisting of twenty thousand to thirty thousand genes. An individual's genome is a complete delineation of their genetic composition. Watson left the NIH because he disagreed with the project's director, who sought patents on sections of DNA

---

15 S. Boggan, "Nobel Winner May Sue over Gay Baby Abortion Claim," *Independent*, October 23, 2011.

16 T. Abate, "Nobel Winner's Theories Raise Uproar in Berkeley: Geneticist's Views Strike Many as Racist, Sexist," SFGATE, November 13, 2000, updated August 6, 2012, sfgate.com/science/article/Nobel-Winner-s-Theories-Raise-Uproar-in-Berk eley-3236584.php.

17 S. Bhattacharya, "Stupidity Should Be Cured, Says DNA Discoverer," *New Scientist*, February 28 2003.

that were being identified. The US Supreme Court has since ruled it is illegitimate to seek patents on DNA that is produced in nature. In 2003, Watson published his own genome online in order to popularize the use of genetic testing. There was one condition: he did not wish to divulge (or even know himself) the status of his APOE gene, which can predict a higher risk for Alzheimer's disease.[18]

Watson's positions at Harvard and the NIH overlapped with his leadership duties at Cold Spring Harbor Laboratory, where he was consecutively the director, president, chancellor, and chancellor emeritus. In 2007, Watson was abruptly relieved of his administrative duties at Cold Spring because of his racist comments in a London interview, wherein he discussed his views that Africans are less intelligent than Westerners and asserted that IQ differences in whites and blacks are due to genetics.[19] He told the *London Times* that while people may like to think that all races are born with equal intelligence, those "who have to deal with black employees find this not true."[20] This was followed by an apology in which he stated he regretted his statements. Watson retired from Cold Spring shortly thereafter, but he was allowed to retain an office there.

Despite his apology, Watson was on a roll and did not restrain himself. In an October 2007 interview with *Esquire*, Watson said, "Some anti-Semitism is justified." In the same interview, he said, "Francis Crick said we should pay poor people not to have children. I think now we're in a terrible situation where we should pay the rich people to have children. If there is any correlation between success and genes, IQ will fall if the successful people don't have children. These are self-obvious facts."[21]

---

18 M. Wade, "Genome of DNA Discoverer Is Deciphered," *New York Times*, June 1, 2007.

19 S. Peck, "James Watson Suspended over Racism Claims," *Telegraph*, October 17, 2007.

20 H. Nugent, "Black People 'Less Intelligent' Scientist Claims," October 17, 2007, *Times*.

21 J. H. Richardson, "James Watson: What I've Learned," *Esquire*, October 19, 2007.

That same year, an analysis of Watson's genome by the company deCODE Genetics showed that 16 percent of his genes are likely to have come from a black ancestor of African descent, suggesting that he had a great-grandparent who was African. Most people of European descent would have no more than 1 percent, so Watson's DNA included sixteen times the number of genes of black origin than found in a typical white European.[22]

The public and professional response to his racist remarks was to cancel all of Watson's scheduled speaking engagements, and since then, he has not been invited anywhere. In 2014, he auctioned off his Nobel medal, citing hard financial times from "having been designated a nonperson."[23] It sold for $4.1 million at Christie's to a Russian billionaire, who returned it to Watson.[24] In 2019, a documentary aired in which Watson reiterated his stance on genetics and race. He said, "And there's a difference on the average between blacks and whites on I.Q. tests. I would say the difference is, it's genetic."[25] At that point Cold Spring Harbor stripped him of his honorary titles and severed all ties; however, on a 2020 scientific publication, Watson still listed an affiliation with Cold Spring Harbor.[26]

Watson also voiced that various mental afflictions are genetic, including schizophrenia, despite the fact that no specific genes for any of the commonly named mental conditions have been identified. Watson's sixteen-year old son Rufus was a former straight-A student

---

22 R. Verkaik, "Revealed: Scientist who Sparked Racism Row Has Black Genes," *Independent*, Monday 10 December 2007.

23 A. Harmon, "James Watson Had a Chance to Salvage His Reputation on Race, He Made Things Worse," *New York Times*, January 1, 2019.

24 I. Sample, "Billionaire Bought James Watson's Nobel Prize Medal In Order to Return It, *Guardian*, December 9, 2014.

25 As quoted in A. Harmon, "James Watson Had a Chance to Salvage His Reputation on Race. He Made Things Worse," *New York Times*, January 1, 2019.

26 G. Bencze, S. Bencze, K. D. Rivera, J. D. Watson, L. Orfi, N. K. Tonks, and D. J. Pappin, "Mito-Oncology Agent: Fermented Extract Suppresses the Warburg Effect, Restores Oxidative Mitochondrial Activity, and Inhibits In Vivo Tumor Growth," *Scientific Reports* 10, no. 1 (2020): 14174.

at an elite boarding school, which he left in October 1985 after being recommended to get counseling. The next month, he ran away and was found threatening to jump from a World Trade Center tower. Rufus was locked up in the psychiatric facility of New York Hospital–Cornell Medical Center in White Plains, from which he briefly escaped six months later.[27] Rufus has been under psychiatric treatment ever since with a diagnosis of schizophrenia, and he remains incapable of independent living.

It is unknown how Watson reconciled his self-professed superiority with the fact of fathering what he must consider to be genetically defective offspring, or how he feels about his own genome data, considering his opinions on the African genetic influence on ability and intelligence.

---

27 M. Aig, "Nobel Prize Winner's Son Found," Associated Press News, May 30, 1986.

13

What Nerve

John Eccles, Alan Hodgkin, and Andrew Huxley were each awarded equal shares of the 1963 Nobel Prize for their various discoveries concerning how nerves signal each other.

John Carew Eccles was born in 1903 in Melbourne, Australia. He became interested in the workings of the brain while studying medicine at the University of Melbourne when he read a book by the neurophysiologist Charles Sherrington.[1] Sherrington had conducted experiments in the late 1890s showing that for every nerve activation there was a corresponding inhibition of an opposing nerve. For example, when the nerves to contract the biceps are fired, the nerves to inhibit the triceps must also be activated, because the biceps can only contract when the triceps are made to relax.

Eccles applied for and won a Rhodes scholarship in order to study at Oxford, with the specific intention of working under Sherrington. Mentored by Sherrington, Eccles's PhD research was on the topic of how impulses get transmitted from nerve to nerve in the central nervous system (the brain and spinal cord). While Eccles was at Oxford, Sherrington's earlier work was recognized by the 1932 Nobel Prize in Medicine, which he shared with Edgar Douglas Adrian.[2]

---

1   C. Sherrington, *The Integrative Action of the Nervous System* (Andesite Press, 2015).
2   *Boneheads & Brainiacs*, 121–23.

Rhodes Scholarship

The Rhodes scholarship was established in the will of Cecil John Rhodes to enable promising young men to study at the University of Oxford, with the goal of encouraging them to become outstanding contributors in the field of public service. Rhodes's will stipulated that it was intended for Germans, colonists of South Africa, and persons from English-speaking countries.

Rhodes was the developer of De Beers Consolidated Mines, through which he owned 90 percent of the world's diamond production by 1891. He then developed the British South African Company, pushing ever farther into native lands to develop diamond mines and exploit cheap labor. Rhodes routinely crafted treaties with various African chiefs that were hardly legal and in any case largely disregarded. As prime minister of the Cape, Rhodes passed legislation that laid the groundwork for apartheid policies, including the Franchise and Ballot Act of 1892, which limited the native vote with financial and educational qualifications, and the Glen Grey Act of 1894, which enforced segregation of native Africans to restricted areas. His policies were summed up by this quote: "The native is to be treated as a child and denied franchise. We must adopt a system of despotism, such as works in India, in our relations with the barbarism of South Africa."[3]

Today, the Rhodes scholarship is open to men and women from all backgrounds and races from around the globe. In 2003 ,the Mandela Rhodes Foundation was established specifically to provide scholarships to African students undertaking postgraduate study in South African universities. Nelson Mandela explained, "We are sometimes still asked by people how we could agree to have our name joined to that of Cecil John Rhodes in this Mandela Rhodes initiative. To us, the answer is easy, and we have explained the logic

---

3   As quoted in B. M. Magubane, *The Making of a Racist State: British Imperialism and the Union of South Africa, 1875–1910* (Trenton, NJ: Africa World Press, 1996).

of our decision on a number of occasions. We have referred to our constitution's injunction for us to come together across the historical divides, to build our country together with a future equally shared by all."[4]

The current version of the Rhodes scholarship website proudly states, "We support all of our Scholars and emphatically affirm their right to speak their minds, protest, mobilise and criticise—including their right to criticise Cecil John Rhodes and his legacy, or the Rhodes Trust itself."[5]

The intricate drawings of the nervous system by 1906 Nobel laureate Ramón y Cajal had revealed there were tiny gaps between individual nerves.[6] Sherrington named these regions synapses, from the Greek *syn*, "together," and *hapsis*, "joining." Meanwhile, and concurrent with Eccles's studies, Henry Dale of Britain and Otto Loewi of Austria demonstrated that chemicals are largely responsible for the transmission of nerve impulses across the synapses between nerves of the peripheral nervous system, that is, the nerves beyond the brain and spinal cord. These discoveries won them the Nobel Prize in 1936, but there remained a great deal of unexplained electrical phenomena in the brain and at other nerve endings.[7] The findings of Dale and Loewi fired the debate between the so-called soupers, who held to the theory of chemical neurotransmitters, and "sparkers" who thought the nerve transmissions were electrical.[8]

---

4   N. Mandela, as quoted by "Mandela Rhodes Foundation," Rhodes Trust, rhodeshouse.ox.ac.uk/about/the-mandela-rhodes-foundation/.

5   "Support for Scholar Protesters and Activists," Rhodes Trust, rhodeshouse.ox.ac. uk/impact-legacy/support-for-scholar-protesters-and-activists/.

6   *Boneheads & Brainiacs*, 141–46.

7   *Boneheads & Brainiacs*, 145.

8   E. Valenstein, *The War of the Soups and the Sparks: The Discovery of Neurotransmitters and the Dispute Over how Nerves Communicate* (New York: Columbia University Press, 2006).

Eccles discovered there was a transfer of sodium, potassium, and chloride across the nerve endings' membranes at the synapses. These electrically charged particles (ions) made an electrochemical gradient that caused a flow of the nerve impulse. Eccles's work showed that an event initiated by chemical neurotransmitters provokes a change in the concentration of ions of sodium and potassium, which in turn promotes electrical flow—and that is how nerve impulses are ultimately conducted.[9] Thus, both parties were correct. It was later found that while most synaptic signaling is launched by chemical neurotransmitters, there are some nerve junctions that are strictly electrical. Eccles further demonstrated that synapses are of different types, with some promoting the transmission of nerve impulses and others that inhibit transmission. He found that a nerve cell is bombarded with signals from many different synapses, and how it responds—firing or not—is determined by which type of synapse predominates.[10]

Eccles married while at Oxford and began a large family productive of nine children. After Oxford, Eccles held teaching and research positions in Sydney and in New Zealand and then settled as professor of physiology at the Australian National University (ANU) in Canberra. His eldest daughter, Rosamund Margaret Eccles, earned her PhD by researching in his neurophysiology labs at ANU and then stayed on to work with her father for about a decade. In 1958 Eccles was honored with knighthood. A few years after winning the Nobel Prize, Eccles remarried, moving to the US and eventually becoming Distinguished Professor of Physiology and Biophysics at State University of New York at Buffalo, where he continued with a robust career in neurophysiology

9   J. C. Eccles, "The Behaviour of Nerve Cells," in *Ciba Foundation Symposium— Nuerological Basis of Behaviour*, ed. G. E. W. Wolstenholm and C. M. O'Connor (London: Churchill, 1958), 28–47.

10 J. C. Eccles, "Excitatory and Inhibitory Synaptic Action," *Annals of the New York Academy of Sciences* 81, no. 2 (1959): 247–64.

research until retirement in 1975.[11] In the 1980s, Eccles played a prominent role in the organization International Physicians for the Prevention of Nuclear War, especially contributing to the scientific and philosophical literature of the movement with such essays as "Planning For Peace,"[12] "Global Cooperation,"[13] and "Global Freedom or Scientific Interchange."[14]

Eccles ultimately produced a stunning 568 publications. Unlike many of his predecessors and contemporaries, who appended their names as authors of papers just because they were in charge of a lab, Eccles's name on a paper invariably indicated his personal participation in all aspects of the investigation. About one-sixth of his publications were philosophical works on the relationship of the brain to the mind and consciousness. In his Nobel banquet speech, Eccles said, "Of course we are still at a primitive level of understanding [of the brain and nervous system]." Furthermore, his discoveries about nerve impulses had not answered the one great question that had dominated his life: "What am I?" Instead he found, "The more we know, the more the mystery grows."[15]

Eccles's penetrating research into the depths of the nervous system led him to reject the materialistic theory that held that the mind of man and his consciousness can be broken down into component

---

11 D. R. Curtis and P. Anderson, "John Carew Eccles 1903–1997," *Historical Records of Australian Science* 13, no.4 (2001): 439–73, science.org.au/fellowship/fellows/biographical-memoirs/john-carew-eccles-1903-1997.

12 J. C. Eccles, "Planning for Peace," *International Seminar on Nuclear War: 3rd Session, The Technical Basis for Peace*, ed. W. S. Newman and S. Stipcich (Singapore: World Scientific Publishing, 1992), 335–40.

13 J. C. Eccles, "Global cooperation," *International Seminar on Nuclear War: 4th Session, the Nuclear Winter and the New Defense Systems, Problems and Perspectives*, ed. A. Zichiichi, W. S. Newman, and S. Stipcich (Singapore: World Scientific, 1992), 434–38.

14 J. C. Eccles, "Global Freedom for Scientific Interchange," *SDI, Computer Simulations, New proposals to Stop the Arms Race: International Seminar on Nuclear War, 5th Session*, ed. A. Zichichi, W. S. Newman, and S. Stipcich (Singapore: World Scientific, 1992), 355–57.

15 J. C. Eccles, Banquet Speech, December 10, 1963, NobelPrize.org, nobelprize.org/prizes/medicine/1963/eccles/speech/.

structures of nerve cells, neurotransmitter chemicals, and electrical impulses. Eccles perceived that the spiritual self is exterior to and the controller of the physical brain. His initial writing on this concept named the controlling mental factor the will. In later publications he called it the mind, and he described the interactions as two-way: brain sensory mechanisms get interpreted by the mind as perceptions, and the mind impinges on the brain to mechanically execute the mind's willed actions.[16] Eccles eventually called the motivating concept the self-conscious mind, defined as implying "knowing that one knows."[17] Consciousness is experienced by this self-conscious mind independent of brain cells, nerve stimulation, or nerve inhibition. Eccles coauthored the 1984 book *The Self and Its Brain* with philosopher Karl Popper, a kindred spirit who helped Eccles express his concepts of consciousness, and it became Eccles's most widely cited philosophical work. The preface of the book remarks on how the two authors' differing religious views did not interfere with their agreements: "One of us (Eccles) is a believer in God and the supernatural, while the other (Popper) may be described as an agnostic. Each of us not only deeply respects the position of the other, but sympathizes with it." In the book, Eccles describes that the self-conscious mind is not identical to some physical part of the cerebral cortex like cells or synapses, although it interacts with the brain to experience the physical world. Simply put, the superior entity of the mind is the interpretative and controlling factor on neural tissue.[18]

John Eccles died in 1997 at the age of ninety-four.

---

16 J. C. Eccles, *The Neurophysiological Basis of Mind: The Principles of Neurophysiology* (Oxford: Clarendon Press, 1952).

17 J. C. Eccles, *The Human Psyche* (Berlin: Springer-Verlag, 1980)

18 J. C. Eccles and K. R. Popper, *The Self and Its Brain: An Argument for Interactionism* (Berlin: Springer International, 1984).

Karl Popper

Sir Karl Raimund Popper was an Austrian-born philosopher who wrote extensively on determinism and free will. He proposed that human rationality was the result of neither random chance nor complete determinism. He more poetically described it as being somewhere between clouds and clocks.[19]

Popper was also a commentator on the philosophy of science. He argued that for a study to be called a true science, it had to have the quality of being able to be subjected to criticism and decisive experiments. For example, Popper rejected psychoanalytical theory as a true science specifically because it was constructed in a way that deflected criticism and prevented testing of its principles.[20] The result of any experiment or rationale that attempted to test the veracity of psychoanalytic doctrines could conveniently be explained away as yet another form of abnormal human behavior.

Alan Lloyd Hodgkin and Andrew Huxley were each awarded one-third of the 1963 prize. They elucidated the exact chemical processes responsible for the passage of impulses along individual nerve fibers. Hodgkin was born in 1914 in England and educated at Trinity College at the University of Cambridge. He was initially the junior teacher for the postdoctoral student Andrew Huxley, who was three years younger.

To study nerves, they used Atlantic squid, which were readily available off the coast of Plymouth, England, where their research station was located. The squid nerve cell was giant and easily manipulated

---

19 K. R. Popper, *Of Clouds and Clocks* (St. Louis, MO: Washington University Press, 1966).

20 K. R. Popper, "Science: Conjectures and Refutations," reprinted in *Philosophy of Science and the Occult*, ed. P. Grim (Albany, NY: State University of New York Press, 1990), 104–10.

in the lab. They pierced it with an electrical probe and recorded what occurred when an electrical stimulus was sent along it. Contrary to the beliefs of the day, which predicted that the electrical impulse would break down the nerve membrane when it reached the nerve cell, they were surprised to find that the impulse caused a rise in the stored-up electrical charge (the "action potential"). They made this discovery in the summer of 1939, and a few weeks later, Hitler invaded Poland, so they just had time to dash off a one-page article on their findings before reporting for duty.[21]

Academic work was suspended while the intellectual resources of the country were turned toward the war effort. Hodgkin initially helped to design oxygen masks for pilots, then for five years, he contributed to the development of much more accurate radar systems. He took scores of test flights to try out and perfect radar modifications, risking at times being shot down when mistakenly identified as an enemy plane.[22] Meanwhile Huxley worked on a small team to integrate radars into analog computers in order to improve the accuracy of nighttime gunnery.

After the war, the team resumed their research by further measuring and trying to explain what caused the findings seen in 1939. It was known that when a nerve cell is at rest, it keeps a large concentration of potassium inside, while keeping most of the sodium outside in the surrounding solution. Hodgkin and Huxley discovered that when an electrical impulse stimulates the nerve cell, the cell responds by letting sodium penetrate its membrane and enter it. This stores up electrical potential and is the basis for nerves "firing"—passing the charge along to the next nerve. Their measuring method and discovery opened up the field that became known as neurophysiology. Today patients get nerve-conduction studies by being pierced with needles that deliver

---

21 A. L. Hodgkin and A. F. Huxley, "Action Potentials Recorded from inside a Nerve Fibre," *Nature* 144 (1939): 710–11.

22 A. L. Hodgkin, *Chance and Design: Reminiscences of Science in Peace and War* (Cambridge: Cambridge University Press, 1992).

and measure electrical charges, facilitating the accurate diagnosis of diabetic neuropathy, compressed nerve roots from a spinal disc, carpal tunnel syndrome, and muscular dystrophy.

Hodgkin served in various research and academic positions, finally becoming Master of Trinity College and was knighted in 1972. Alan Hodgkin died in 1998 at the age of eighty-four.

Master of Trinity College

Trinity College is one of thirty-one colleges within the University of Cambridge. Trinity was founded by King Henry VIII in 1546 when he combined several smaller academic institutions. The Mastership of Trinity is a Crown appointment made by the English monarch on the recommendation of the college. The Master is responsible for superintending the running of the college.

Historically, the Master must have held a degree from Cambridge and was usually a member of Trinity College staff. For the first few centuries, Trinity Masters were often church figures, including clergymen, preachers, chaplains, deacons, priests, or bishops. The past Masters, of the clergy or not, have had diverse areas of accomplishment, with some being playwrights, mathematicians, naturalists, philosophers, historians, poets, physicists, authors, physiologists, linguists, economists, astrophysicists, biologists, and at least one a political-party leader. The current (and thirty-ninth) Master is a woman physician who did not graduate from Cambridge.[23]

Andrew Huxley was born in 1917 and was a postgraduate student at Trinity with no previous research experience when Hodgkin mentored him at the Plymouth research station in 1939. Huxley's unique

---

23 "The Master of Trinity," Trinity College Cambridge, trin.cam.ac.uk/fellows/master/.

contribution was to devise the mathematical equations that provided a rational explanation for the electrical observations the pair studied. He was still fairly early in his career when they conducted the follow-up experiments in the postwar years that would earn them recognition with a Nobel Prize years later.[24]

Huxley made another groundbreaking discovery in the early 1950s while collaborating with the German physiologist Rolf Niedergerke. Huxley developed a new light microscope that could study living muscle fibers. This made it possible to observe how muscles contract, specifically by visualizing what happened to the protein filaments that attach individual muscle fibers. Huxley and Niedergerke saw that the filaments changed in length as the muscle bunched up, and they discovered that the filaments glided across each other in a sliding motion, creating muscle shortening (contraction).[25] Again, Huxley's unique contribution was to devise the mathematical equations that explained what they saw. These equations have subsequently been applied to movement studies of all animals above the level of bacteria.

The Famous Huxleys

Andrew Huxley was born in 1917 into the famous Huxley family. His grandfather Thomas Henry Huxley was a biologist who was a contemporary of Charles Darwin. While Darwin was examining sea turtles in the Galapagos, T. H. Huxley was studying crayfish in the Great Barrier Reef and off the southern coast of New Guinea. Both men observed consistent physical traits that linked various species, suggesting what later came to be known as the theory of evolution. Darwin's 1859 book, *On the Origin of Species,* introduced the

---

24 A. F. Huxley, "The Quantitative Analysis of Excitation and Conduction in Nerve," December 11, 1963, NobelPrize.org, nobelprize.org/prizes/medicine/1963/huxley/lecture/.

25 A. F. Huxley and R. Niedergerke, "Structural Changes in Muscle during Contraction; Interference Microscopy of Living Muscle Fibres," *Nature* 173, no. 4412 (1954): 971–73.

concept of evolving, generation by generation, by the process of natural selection.[26] The book provoked widespread debate for its social and religious implications, although Darwin was a religious man and did not find that his theory denied a prime role for a divine creator. On the other hand, T. H. Huxley was a dedicated agnostic, a term he coined in 1869 to describe that one can never know the ultimate reality of material worlds or spiritual worlds. T. H. Huxley recognized that many critics had not read or understood Darwin's book, and he sought to clarify matters by giving a series of "Lectures to Working Men."[27] He later became an effective administrator for organizing and elevating the status of science and technical education in England.

One of Andrew's half brothers was philosopher Aldous Huxley, author of *Brave New World*, a science fiction work published in 1932 that described a horrific totalitarian society based on human cloning five hundred years in the future.[28] Later in life, Aldous popularized self-exploration through psychedelic substances and was instrumental in what became the Human Potential Movement. After a struggle with throat cancer, Aldous ended his life by having his wife inject him with two shots of LSD.[29]

Another half brother was Julian S. Huxley, an evolutionary biologist and eugenicist, who served as vice president and then president of the British Eugenics Society. Julian advocated for social policies that would gradually eliminate the lowest classes of people, whom he considered genetically defective. He advised that the poor should not be given easy access to health care—leaving them

---

26 C. Darwin, *On the Origin of Species by Means of Natural Selection, or The Preservation of Favoured Races in the Struggle for Life* (London: John Murray, 1859).

27 T. H. Huxley, *The Present Condition Of Organic Nature* (n.p., 1863; Project Guetenberg, 2013, gutenberg.org/files/2921/2921-h/2921-h.htm#link2H_4_0001).

28 A. L. Huxley, *Brave New World* (New York: Harper & Brothers, 1932).

29 L. Huxley, *This Timeless Moment: A Personal View of Aldous Huxley* (New York: Farrar, Straus & Giroux, 1968).

> to die of untreated diseases would allow natural selection to take its course.[30]

In a 2001 video interview at the Nobel Foundation, Andrew Huxley spoke of how popular trends in science promote disregard for earlier discoveries and even the suppression of ideas that are not riding at the front of scientific fashion of the day. He gave the example of his own Nobel Prize–winning research, which was accomplished in part by using an ordinary light microscope to visualize nerve cells. At that time, he and Hodgkin were unaware of the 1902 paper by Charles Ernest Overton that had documented nearly the same findings on muscle cells.[31] (Overton died in 1933 without his work being recognized.) Huxley attributed the oversight to the fact that at the time of Overton's discovery the rising star in science was molecular biology. This popular scientific trend dictated that all mysteries of life would be explained by molecules, and that there was no room for light microscopes anymore—so why bother with them? Writing eighty years later, a science historian suggested that the original discovery of action potentials was made even earlier, in 1862, by Julius Bernstein, a discovery that was overshadowed by the rise of germ theory at that time.[32]

Then it happened again: Long after his work on muscle contraction, Huxley belatedly found out that the sliding-muscle-fiber theory had originally been postulated by a German physiologist in 1870.[33]

---

30 J. S. Huxley, *Man in the Modern World* (London: Chatto & Windus, 1947).

31 E. Overton, "Beiträge zur allgemeinen Muskel- und Nerven-Physiologie. II. Über die Unentbehrlichkeit von Natrium- (oder Lithium-) Ionen für den Contractionsact des Muskels" [Contributions to general muscle and nerve physiology. II. about the indispensability of sodium (or lithium) ions for the contraction act of the muscle], *Pflügers Archiv* 92: 346–86.

32 S. Schuetze, "The Discovery of the Action Potential," *Trends in Neurosciences* 6, no. 5 (1983): 164–168.

33 A. F. Huxley, Interview, December 5, 2001, NobelPrize.org. nobelprize.org/prizes/medicine/1963/huxley/interview/.

In another example, Huxley was well aware of the controversy surrounding Darwin's theory of natural selection that his famous grandfather had so publicly defended, earning the elder Huxley the nickname of "Darwin's bulldog." By 1900 natural selection was a well-accepted scientific theory. In just a couple of decades, that secure position came undone when it was found that chromosomes underwent chance mutations. Suddenly, the position swung 180 degrees to the idea that all biological changes happened by sheer chance rather than by steady, gradual natural selection for survival of the fittest. In fact, both processes are continually happening.

Andrew Huxley observed that when popular fashion authorizes one scientific theory to the exclusion of all others, it is damaging to scientific creativity and constructive debate. To remedy the problem, Huxley advocated much stronger education in the specifics of science history.

Huxley had a distinguished academic career, and like Hodgkin, he was knighted. He took over as Master of Trinity College when Hodgkin stepped down and held the position until his death in 2012.

14

The Cholesterol Discoveries

Konrad Bloch and Feodor Lynen won the 1964 Nobel Prize for their discoveries concerning the mechanism and regulation of cholesterol and other fats.

Feodor Lynen was born in 1911 in Munich, Germany, where he was educated and spent most of his research career. As a young boy he tried chemistry experiments at home, once causing an explosion and ruining his best trousers. After that, he left the subject alone until getting formal training.

Lynen earned his PhD by studying the toxic substances of *Amanita*, a family of mostly poisonous mushrooms. He then switched his attention to a more acceptably intoxicating substance, beer—actually, he was studying yeast, which he obtained from the Löwenbräu brewery in Munich. The yeast were convenient lab specimens in which to study biological reactions.

Amanita Mushrooms

Although some species are edible, *Amanita* are responsible for the majority of deaths worldwide from poisonous mushrooms. The so-called death cap mushrooms are reported by survivors to be quite tasty. Symptoms of poisoning don't occur for two or three days after eating, but just one-half of a typical death cap mushroom is enough to kill an adult man. Charles VI of the Holy Roman Empire is presumed to have died ten days following a sautéed mushroom dish, with his symptoms suggestive of death cap poisoning. The related destroying angel mushroom causes symptoms eight to twenty-four hours after eating, and 60–80 percent of victims do not survive. The panther cap mushroom causes a speedy sensation and hallucinations, which can progress to coma and death. Fly agaric is another poisonous *Amanita* mushroom that can be deadly when overdosed; however, it has also been used in minute amounts in shamanic rituals for its hallucinogenic properties.

Lynen utilized culture broths of yeast to study the molecular processes of metabolism. Others had already discovered some aspects of how a cell metabolizes the components of food to turn it into energy for living, but how the cell breaks down fat and makes cholesterol was still unknown.

Lynen's research was hampered during WWII by the difficulties in getting lab equipment, by Nazi policies that curtailed communications with international researchers, and finally by the bombing of Munich. After the war, Lynen had to reapply for his position as professor of chemistry while his character, credentials, and wartime activities were carefully evaluated pursuant to Article 131, which facilitated the reemployment of persons who had served the German Reich (such as public servants and university professors) but had not been classified as Nazis during the postwar denazification process. Passing scrutiny came to be known as receiving the Persilschein, or Persil, certificate, named after a popular brand of laundry powder, implying that one now was recognized as ethically clean. Although Lynen had kept his university position throughout the Nazi era, he was never involved in work that was used toward the war effort, the holocaust, or related activities of the Third Reich.

Lynen's breakthrough came in 1951, with research demonstrating the key first step in a chain of reactions that break down fatty acids. Fatty acids are the beginning point for the building of more complex structures that form the walls of every cell in the body.

Fatty acids also attach to proteins to regulate their activity. Another significant role of fatty acids is to serve as an energy storage system that can be tapped as a biological source of fuel when more readily consumable carbohydrates become scarce.

Specifically, Lynen sought to discover the structure of acetyl coenzyme A, something he and others around the world had been working on for years. It is the starting molecule of all fatty acid and cholesterol synthesis. When Lynen finally discerned the exact arrangement of the molecules in the substance, he saw that it was incredibly simple— so simple that he was profoundly worried that someone else would hit upon the idea in the weeks it took for his findings to be published in a journal of the German Chemical Society.[1] No one trumped him, and the paper instantly brought Lynen international attention in the world of biochemistry.

There followed another decade of experiments on how fatty acids and cholesterol are formed, degraded, and distributed in the cells and throughout the body. In 1957, researchers under Lynen's supervision discovered a key intermediate compound in parallel with the same discovery in Konrad Bloch's lab in America. Both men recognized that these discoveries offered opportunities to make medications that would target one or another metabolic steps to lower cholesterol that had gotten too high.[2]

Lynen enjoyed an active lifestyle despite limping due to a locked knee from breaking his leg twice in ski accidents. He would regularly

---

1   F. Lynen et al., "Zum biologischen Abbau der Essigsäure Vol1. 'Aktivierte Essigsäure,' ihre Isolierung aus Hefe und ihre chemische Nature" [The activated vinegar—their isolation from yeast and their nature], *Annalen der Chemie* 574, no. 1 (1951).

2   F. Lynen, "Cholesterol und Arteriosklerose" [Cholesterol and atherosclerosis], *Naturwissenschaftliche Rundschau* 25: 382–387.

take hikes in which he was sure to encounter a tavern, for according to his biographers, "Enough alcohol to create a carefree and jolly atmosphere was for him an essential part of life."[3]

On the occasion of his sixty-fifth birthday, former students and colleagues presented written memories of their time in Lynen's lab, and the picture emerges of a kind and fun-loving man who was "obviously in love with life and whose capacity to enjoy it in terms of hearty food and good potables (till all hours of the night!) seemed gargantuan."[4] Another fond recollection describes "this human dynamo of a man with sparkling eyes, a charming smile and unsurpassed endurance, who even after a night of boisterous festivity was always able the next day to appear bright-eyed and smiling, and give a brilliant lecture."[5]

Lynen was in the habit of directly conducting experiments to test plausible ideas instead of wasting time deliberating the pros and cons of maybe doing an experiment. His experimental designs were described as genius, elegant, and simple.[6] He demanded that his students persevere in their courses of investigation, lest they miss out on a discovery just because of giving up too soon. He did not keep any one junior investigator in his lab for longer than a few years, considering that if the student were dull, he needed to go, and if the student were bright then he must have his own ideas, so he also needed to go—to start his own lab! Lynen demanded a lot of his students over the span of his thirty-seven-year career in academia, but never more than he required of himself. "To me academic freedom means that one is allowed to work longer hours than the rules require," he said.[7]

---

3   H. Krebs and K. Decker, "Feodor Lynen, 6 April 1911–6 August 1979," *Biographical Memoirs of the Royal Society* 28 (November 1, 1982).

4   G. Hartmann, ed., *Die aktivierte Essigsdure und ihre Folgen* [The activated vinegar and its consequences] (Berlin: DeGruyter, 1976).

5   H. G. Wood, "Obituary: Feodor (Fitzi) Lynen," *Trends in Biochemical Sciences* 4, no. 12 (1979), doi.org/10.1016/0968-0004(79)90293-7.

6   E. Maier, "Feodor Lynen: The Great Experimenter," Max Planck Gesellschaft, April 05, 2011, mpg.de/1344155/Feodor_Lynen.

7   As quoted in H. Krebs and K. Decker, "Feodor Lynen, 6 April 1911–6 August 1979," *Biographical Memoirs of the Royal Society* 28 (November 1, 1982).

Konrad Bloch was born in 1912 to Jewish parents in Neisse, then located in Prussia and now within the boundaries of Poland. Like Lynen, Bloch was educated in Munich, where he studied under the freshly minted Chemistry Nobel Prize winner Hans Fischer. After completing the equivalent of a master's degree in chemistry, Bloch applied for his doctoral work but was notified by the dean of the school that Dr. Fischer had turned him down. Bloch later learned this was a lie; his appointment was declined because of the new racial laws meant to scrub public institutions of those with any Jewish heritage.

Bloch next applied to the university in Danzig. The 1919 Treaty of Versailles had made this largely ethnically German town into a "free city" with administrative ties to Poland. Bloch was lucky to be turned down there, too, as Hitler's army would occupy Danzig and then all of Poland in 1939. Bloch tried his luck at Utrecht University in the Netherlands and was—again fortunately—declined. Soon after, the Nazis moved into Holland. Bloch was finally accepted into the labs at the Swiss Research Institute at Davos, in neutral Switzerland. There he studied the fats present in the tubercle bacterium. His two papers on this topic were presented for his PhD thesis but rejected simply because one member of the committee was offended that Bloch's thesis did not make reference to the committee member's own publications on a related topic. By then it was 1937. Bloch was at a dead end academically at Davos and trapped in Europe.

He applied for and got a letter of acceptance to continue his PhD work at Yale, which was accompanied by another letter informing him that the position would be unpaid. The acceptance letter secured him safe passage on a ship out of Europe with only the clothes on his back. Upon arrival in America, Bloch applied for a paid graduate position at Columbia in New York.

At Columbia, Bloch submitted the same two papers describing his earlier work, which were accepted as partial contributions to his doctoral thesis, but Columbia insisted he had to do something there in order to earn a degree from the institution. He was able to complete a small biochemical project that had stumped other researchers, and his PhD was finally awarded in 1938. In a nicely paid postdoctoral-research position at Columbia, Bloch worked under Rudolf Schoenheimer, who assigned him to cholesterol research. Schoenheimer committed suicide in 1941, leaving Bloch to become the country's premier researcher in the study of cholesterol.[8]

Cholesterol is necessary to make the structure of the walls of every cell in the body. It insulates nerve fibers to speed the transmission of impulses, and it is the most prevalent organic molecule in the brain of humans. It is the basic building block of vitamin D, cortisol, and all sex hormones, including estradiol, progesterone, and testosterone. Cholesterol helps to absorb fats from food as well as the fat-soluble vitamins like vitamins A, E, and K. It was an important molecule to study from a health perspective, too, since some genetic variants can cause abnormal accumulations of cholesterol, leading to an increased risk of heart attack and stroke. Bloch continued his work at the University of Chicago and then at Harvard, where he focused on the synthesis of fatty acids and cholesterol, while Lynen's work focused on their breakdown.

Blonde Venetians

Konrad Bloch had fun applying biochemistry to problems in real life. He became intrigued about the fake blondes so prevalent in Venetian paintings from the fifteenth and sixteenth centuries. These are exemplified in the works of Renaissance painter Tiziano Vecelli (or Vecellio), also known in English as Titian, such as his *Venus and Her Mirror*. Bloch found a variety of different archaic ingredient lists for a hair rinse called *aqua bionda*. He could not locate exact

8   M. F. Shaughnessy, "An Interview with Professor Manuel Varela: 'Konrad Emil Bloch—What does he have to do with Cholesterol?,'" Education Views, July 4, 2019, at   educationviews.org/an-interview-with-professor-manuel-varela-konrad-emil-bloch-what-does-he-have-to-do-with-cholesterol/.

directions, but they all required extensive sun exposure on the treated hair. The lists had in common some plants, such as boxwood, cypress blossoms, cumin, myrrh, clover, and the dregs of white grapes. Most ingredient lists also included saltpeter (potassium nitrate) and alum (aluminum). Bloch supposed that the dyeing process somehow generated hydrogen peroxide, which is the basis of modern hair dyes. Hydrogen peroxide was not discovered in a chemistry lab until 1818 and was not detected in nature until 1888, when it was found in rainwater. Bloch developed theories and even proposed biochemistry experiments to test how combinations of the common ingredients in aqua bionda formulations would generate hydrogen peroxide when exposed to sunlight. However, he never conducted the experiments. Maybe someday the National Institutes of Health will fund a doctoral student to solve the mystery of the blonde Venetians![9]

The discoveries of Lynen and Bloch laid the foundation for decades of subsequent experimentation to find a natural substance or a drug to limit the manufacture of cholesterol in the body. In the 1970s, the Japanese microbiologist Akira Endo was fermenting a broth of penicillium mold in an effort to discover new antibiotics when he incidentally discovered something that inhibited the key step in cholesterol synthesis. He was able to isolate the exact substance. Animal tests showed it was too toxic to try on humans, but biochemists at the pharmaceutical company Merck pursued the line of research with another mold, eventually finding the same substance in a different broth, this one with lower toxicity. This became the first so-called statin, a group of cholesterol-lowering drugs. The first statin became FDA approved

---

9   K. Bloch, *Blondes in Venetian Paintings, the Nine-Banded Armadillo, and Other Essays in Biochemistry* (New Haven, CT: Yale University Press, 1997).

in 1987, and within two decades some 40 percent of adult Americans were taking a statin.

The movement toward treating high cholesterol has been questioned because of a critical reappraisal of the statin manufacturer's initial studies: they were found to have less than impressive results. Furthermore, cholesterol was once identified as a major risk factor for cardiovascular disease, but it is now recognized as a minor risk factor. In addition, there is increasing appreciation for the beneficial effects of cholesterol naturally produced in the body, which include lining the nerves in the brain to facilitate speedy impulse transmission. This may be adversely affected by excessively lowering cholesterol with statins.

Feodor Lynen died in 1979 from complications of surgery to repair an abnormally dilated section of his abdominal aorta. The condition is due to atherosclerosis, or hardening of the arteries, which paradoxically weakens the artery wall and can lead to dangerous dilation and rupture. Lynen's condition occurred before the marketing of cholesterol-lowering drugs. Konrad Bloch died in 2000 from congestive heart failure, but it is unknown if his particular condition was related to high cholesterol.

15

The French Freedom Fighters

The 1965 prize was awarded to André Lwoff, François Jacob, and Jacques Monod for their discoveries concerning genetic control of enzymes and viruses. This French trio was much more than a collegial group of molecular biologists, for they all served to free France from the German occupation during WWII.

Andre Michel Lwoff was born in France in 1902 and was educated at the University of Paris, joining the Pasteur Institute at the age of nineteen. He earned his medical degree and PhD, and then studied briefly in Germany and in the US under Rockefeller grants.

Lwoff was preceded at the Pasteur Institute by Félix d'Hérelle, who proposed the idea that bacteria could harbor viruses in either a dormant state, when they were not replicating or harming the host bacteria, or in an actively replicating state, eventually disintegrating the bacterial cell.[1] This concept was to develop into Lwoff's Nobel Prize–winning research. By the time Lwoff joined the Pasteur Institute, his senior colleagues were Eugène and Elisabeth Wollman, who encouraged him to join them in studying the phases of viral life inside bacteria.[2]

---

1   F. D'Herelle, "Sur un Microbe Invisible Antagoniste des Bacilles Dysentériques" [An invisible microbe that is antagonistic to the dysentery bacillus], *Comptes Rendus Hebdomadaires des Séances de l'Académie des Sciences* 65 (1917): 373–75.

2   E. Wollman and Elisabeth Wollman, "Les 'phases' des bactériophages (facteurs lysogenes)" [The phases of bacteriophages], *Comptes Rendus Hebdomadaires des Séances et Mémoires de la Société de Biologie* 124 (1937): 931–34.

Pastorians

Louis Pasteur, a chemist and biologist, founded the Pasteur Institute in Paris and was the first to give scientific evidence for germ theory, which held that organisms too small to be seen were the causes of many diseases. By the end of the nineteenth century, germ theory was rapidly replacing the two other prevailing beliefs: miasma theory, which attributed infectious diseases to some ill-defined menace in the air, and the theory of spontaneous generation, which presumed that infectious agents could arise out of inanimate objects or dead tissue without having living predecessors—like lice growing out of dust or maggots suddenly appearing on dead flesh. Pasteur developed the first modern and effective vaccine for rabies and saved the silk industry by discovering the cause of silkworm disease. However, his most famous discoveries emerged from his research on behalf of vintners and brewers who wanted to know why their wines and beer sometimes went sour. He found that bacterial overgrowth was responsible for the brew gone bad. He recommended heating the beverages to just below boiling to kill off most of the bad bacteria. The process was named pasteurization, and it preserved the taste of beer and wine while preventing spoilage. Most important to the vendors, pasteurization prolonged shelf life. First-generation Pastorians were colleagues and students of Louis Pasteur who fervently supported his ideas about the cause of infections, and they contributed to and furthered Pasteur's research. Subsequently, graduates of the Pasteur Institute were often called "disciples," spreading the use of scientific methods to locate and identify germs, building on the methods established by their spiritual and scientific father, Louis Pasteur, who died in 1895.

Lwoff became the head of the Pasteur Institute in 1938. He and the Wollmans continued their virus research even as France was occupied by the Nazis, despite the risk to the Wollmans, who were Jewish.

In December 1943, Elisabeth Wollman was arrested at home, and Eugène Wollman was seized at the Pasteur Hospital, where he was convalescing from an illness. They were transported to Auschwitz and murdered upon arrival. In 1940, their son Elie Wollman had illegally crossed into the so-called Free French Zone where he worked under an assumed name as a physician to French Resistance operatives. In Paris, Lwoff continued his research as best he could while serving the French Resistance by gathering intelligence for the Allies. Lwoff also hid American airmen in his apartment after they had been shot down over Nazi-occupied territory.

Free French Zone: *Zone Libre*

When Germany occupied France in June of 1940, they allowed the southeastern portion of the country to be administered by a nominally French government headed by WWI hero Philippe Pétain and largely run by his minister of state, Pierre Laval. Based in the city of Vichy, this fully collaborationist government enthusiastically emulated Nazi policies and contributed to Nazi propaganda. Germany took full possession of the zone in November 1942, which positioned them closer to the threatening Allied forces in northern Africa. Pétain and Laval continued to aggressively administer the Nazi programs of oppression and genocide. When the Allies liberated France in 1944, Pétain, Laval, and the French cabinet fled to Germany for protection. After the war, Pétain was convicted of treason, but due to his advanced age (eighty-eight) his death sentence was commuted to life in prison. He died at the age of ninety-five. After the war, Laval escaped Germany to Spain, from which he was expelled, then fled to Austria, where he finally surrendered to the Americans. He was condemned to death for treason. After a suicide attempt, he was nursed back to health so that he could be executed in October 1945.

After the war, Lwoff was awarded the Médaille de la Résistance by Charles de Gaulle. In 1946, Elie Wollman came to study at the Pasteur Institute, where he eventually became Lwoff's assistant on the research of viruses that infect bacteria, which were called bacteriophages. They conducted painstaking experiments to demonstrate the existence of two pathways that an infecting virus can take: A virus can immediately use the bacteria's machinery to replicate in great numbers, quickly bursting the host cell and releasing virus copies to infect nearby bacteria. On the alternate pathway, the infecting virus does not cause its immediate host to die; instead it hacks into the bacterium's genes to incorporate its own viral gene. There it does no harm to the host organism, at least not immediately.

Lwoff, Elie Wollman, and their many scientific collaborators further demonstrated that the incorporated viral gene gets passed on to subsequent bacterial generations during cell division by hitching a ride within the host's DNA. Only in later generations does the viral gene become active, usually after ten thousand bacterial cell divisions, or sooner if some stress—like ultraviolet radiation, for example—is applied. At that time, the viral gene takes the other path, using the bacterial cell machinery to replicate, ultimately disintegrating the host cell, and spreading new virus particles to infect more cells.

When these bacteriophages (literally, "bacteria eaters") are passed down to succeeding bacterial generations in a noninfective form, they are called prophages. The process was named lysogeny (from *lyso*, "breaking open," and *geneia*, "producing or generating").[3] Other terms applied to the players in this microdrama are amusingly anthropomorphic. The prophages are also called temperate bacteriophages, in the sense that they are restrained guests—just being polite, while surreptitiously ingratiating themselves into the very fabric of the host DNA. The actively replicating viruses are called virulent bacteriophages, in

---

3   V. Racine, "Lysogenic Bacteria as an Experimental Model at the Pasteur Institute (1915–1965)," Embryo Project Encyclopedia, October 10, 2014, embryo.asu.edu/pages/lysogenic-bacteria-experimental-model-pasteur-institute-1915-1965.

the sense of poisonous or evil, capable of destruction. After acting like very well-behaved permanent guests for many bacterial generations, they show their true intent. The experimental work of Lwoff and colleagues demonstrated that an acquired characteristic, such as a viral gene, can be subsequently inherited, since it is carried within the host's genes. Only generations down the line does it become virulent.[4]

Lwoff described a viral infection as "the introduction into a cell of the genetic material of a virus." This did not always mean death for the infected host cell. "When the infected cell survives, the infection is said to be integrative. Two 'parts,' the cell and the virus, have been united into a whole: this is an integration. A new system has been formed. The new system, the cell/virus system which behaves and reproduces as a whole, is different from the original cell, for the 'infected' cell carries and perpetuates the genetic material of the infecting virus."[5]

Etymology of *Virus*

The first known use of the word *virus* was in 1599, in its archaic meaning of "poisonous fluid, or venom," from Latin. The use of *virus* in science goes back to 1898, when the Dutch microbiologist Martinus Beijerinck was reporting on the cause of a contagious disease that left blotches on tobacco leaves. He described an infectious substance that was able to be passed through a filter that had pore sizes small enough to trap bacteria. Not conceiving of a particle smaller than bacteria, he thought the infectious substance was the fluid itself. His report called it a contagious living liquid, or "virus." [6] The infectious property was subsequently determined

---

4   A. Lwoff, "Lysogeny," *Bacteriology Review* 17 (1953): 269–337.

5   A. Lwoff, "Tumor Viruses and the Cancer Problem: A Summation of the Conference," *Cancer Research* 20 (1960): 821.

6   M. Beijerinck, "Ueber ein Contagium vivum fluidum als Ursache der Fleckenkrankheit der Tabaksblätter" [Concerning a contagious living fluid as the cause of the blotch disease on tobacco leaves], *Discourses of the Royal Academy of Sciences in Amsterdam*, Second Section, VI, no. 5 (1898).

to be a characteristic of discrete physical entities called "filterable particles." This was such an ungainly phrase that the word virus was globally adopted. A virus was first visualized as a distinct particle by use of the electron microscope in 1939.

Lwoff's discoveries clarified that viruses cannot exist alone, and so they are parasites within a cell. He explained in his Nobel lecture: "Here it finds whatever it lacks: enzymes, building blocks, a source of energy, and ribosomes. The virus is necessarily an intracellular parasite."[7] These discoveries immediately gave rise to questions of what regulates the timing of these viral and bacterial affairs.

———

Throughout the period of Lwoff's research, fellow Pasteur researcher Jacques Monod had been studying a parallel problem of regulation in bacteria: What regulates the genes that code for the bacterial cell to make enzymes that digests the food it consumes? Monod's work led to the answers.

Monod had long studied the control of the enzymes employed by bacteria to digest sugars. He had observed that bacteria prefer to consume glucose, and their genes readily code for enzymes to use glucose as an energy source. He learned that when all available glucose was depleted, it took a while before the bacteria ramped up production of a different enzyme that was capable of utilizing the less-preferred sugar fructose. While looking for a biochemical "on" switch that permitted previously dormant genes to become expressed and go into action on the substitute sugar, Monod found a bit more complexity: In fact, the gene that codes for the second enzyme (for digesting fructose) is always there and ready to go, but an immediately neighboring section of gene is constantly inhibiting it. Only when there is none of the preferred

7   A. Lwoff, "Interaction among Virus, Cell, and Organism," December 11, 1963, NobelPrize.org, nobelprize.org/prizes/medicine/1965/lwoff/lecture/.

glucose sugar will the inhibition get turned "off," thus allowing for the gene to freely code for the enzyme to digest the fructose sugar.

Jacques Monod was born in 1910 to an American mother from Milwaukee, Wisconsin, and a French father who was a painter. Monod studied biology at the Sorbonne and was in the midst of PhD work there when he interrupted his studies to join the French military in anticipation of German hostilities. In May of 1940, the Nazis overran the poorly planned and overconfident French defenses. The French military was ordered to disband, as did Monod's unit. He returned to occupied Paris to resume studies at the Sorbonne. Monod and his wife, Odette, the granddaughter of a former high rabbi of Paris, were required to register with the Nazi authorities, and she had to wear a yellow star marking her Jewish heritage. At the end of 1941, Monod joined a communist-inspired French Resistance unit, although he was never a communist. He wrote propaganda tracts and recruited fellow students to the free French cause. In 1942, the Nazis started a series of roundups of Jewish citizens who were immediately murdered, perished en route, or killed in concentration camps. Odette stopped wearing her yellow star, changed her hair color, and obtained false papers to relocate with the couple's twin children to the Free French Zone for the duration of the war.

Meanwhile, the Sorbonne became too dangerous to inhabit because of crackdowns by the German occupying forces, so Monod continued his research in a small attic laboratory at the Pasteur Institute. Early on, he sustained a raid of his lab by the German military police in which they found no evidence of his connection with the underground. As the war progressed, the seizure and torture of Resistance workers became increasingly common, so there was always a grave threat that Monod's identity would be revealed by captured comrades who were questioned under torture. Monod was forced to abandon his research to go fully underground, including disguising his physical appearance.

Monod eventually became the chief of staff of the overarching Resistance organization called French Forces of the Interior, in which post he directed sabotage, such as the bombings of supply factories, fuel depots, and rail lines. He arranged for parachute drops of weapons and the interception of essential German mail. Monod also coordinated the actions of disparate resistance factions and orchestrated radio networks in order to facilitate communication with Allied agents.

After the war, Monod was awarded the Croix de Guerre and the American Bronze Star Medal and was honored with the rank of Chevalier in the Légion d'Honneur. He also sporadically continued activist activities. He made several attempts to assist the escape of Hungarian scientists Agnes Ullmann and her spouse Tamas Erdos, who were being closely monitored by the occupying Soviets for their connections with the failed Hungarian revolution. Monod was finally successful in helping arrange the smuggling of the couple through the tightly controlled Hungarian-Austrian border by hiding them in a compartment under a laundry box in a camper van. They narrowly escaped detection when the camper was boarded and carefully inspected by Soviet authorities. Agnes Ullmann subsequently had a distinguished scientific career at the Pasteur Institute, eventually becoming its laboratory director.

———

It was François Jacob, the third prizewinner, who saw how the parallel tracks of Lwoff's and Monod's research could merge to investigate exactly how viruses are regulated to be either immediately virulent (lytic) or to remain dormant for many bacterial generations (lysogenic).

Jacob was born in France in 1920, and he was in medical school when WWII erupted, which forced an abrupt change of course. He heard an illegal radio broadcast by Charles de Gaulle, the French military leader who had established temporary command in England to join the Allies in fighting for France. De Gaulle urged eligible French men to quickly assemble in England to join his army. Jacob and a friend pushed through hundreds of people desperately trying to get passage out of France, managing to board a transport boat moving Polish troops to England. Despite his scant training, Jacob was assigned as a medical officer. Over the next four years, he served in campaigns fighting the German attacks in England, Senegal, the French Congo, Gabon, Cameroon, Chad, Libya, and Tunisia.

Finally, in the summer of 1944 his fully French unit landed at Normandy on the heels of the massive Allied D Day invasion. The plan was to allow the French to be the first to march on Paris to drive the Nazis out. Jacob was not to join them, because when they were yet two hundred miles short of the city, their slowly advancing column suffered a German air attack. He bolted from his tank and dove into a ditch, but a friend caught enemy fire and lay out in the open. Jacob crawled to his aid, but found his friend too wounded to move, though alert enough to beg Jacob not to abandon him. When another air attack swooped down, Jacob huddled against his friend. Jacob sustained more than fifty shell fragments, fracturing his arm, leg, and pelvis. His buddy died in in field. Jacob was taken to a hospital while the remainder of his unit advanced to free the city of Paris.

After a slow recuperation, Jacob completed medical school but realized his physical impairments would preclude the traditional practice of medicine. He submitted his very poor resumé three times to work at the Pasteur Institute as a medical school graduate with an interest in microbiology but virtually no experience. Upon finally gaining acceptance, he was informed that he would be assigned to "phage work," which he promptly headed to the library to look up. The term was too new to be found in any book, and he had no idea what he had signed up for.

Along with Elie Wollman, Jacob and Monod used lysogenic bacteria to conduct hundreds of experiments in an effort to find what determines how and when the quietly carried bacteriophage genetic material transforms into deadly replicating virus. They detected the existence of a regulatory gene that was immediately adjacent to the viral gene section that had hacked into the bacterial DNA. The function of the regulatory gene is to modulate the activity of inhibitors that suppress the neighboring DNA code from being translated into certain enzymes. By suppressing the enzymes, the virus portion of the DNA was suppressed from leaping into massive replication.[8]

These findings were later crystallized into a concept of a hierarchy of genes: there are sections of the DNA that code for enzymes, and there are adjacent sections that repress the code next door. Jacob and Monod concluded that very specific sections of DNA code for regulatory proteins that act as "on/off switches," letting some genes be expressed or not, controlling the life cycle of viruses within the bacterium.[9] They predicted that this fundamental mechanism would be discovered in the cells of all living organisms, which has since been found to be true. As Monod said, "What is true for E. coli is true for the elephant."[10]

All three Nobel winners continued with productive research. Lwoff discovered that vitamins serve both as growth factors for microbes and as necessary coupling agents with enzymes (coenzymes). Monod contributed to the expanding body of knowledge of enzyme control. Jacob pioneered ways to map the locations of specific genes on bacterial chromosomes and on phages.

The more they discovered, the more there was to know. Toward the end of his career, Jacob observed, "The main discovery during this century of research and science has probably been the depth of our ignorance of nature. . . . For a long time, we claimed to understand how things worked. Or we simply made up stories to plug the gaps."

The three banded together from time to time for social causes, including becoming vocal proponents of reproductive rights in France,

---

8 F. Jacob and J. Monod, "Gènes de structure et gènes de regulation dans la biosynthèse des proteins" [Genes of structure and genes of regulation in the biosynthesis of proteins], *Comptes Rendus Hebdomadaires des Séances de l'Académie des Sciences* 249 (1959): 1282–84.

9 F. Jacob and J. Monod, "Genetic Regulatory Mechanisms in the Synthesis of Proteins," *Journal of Molecular Biology* 3 (1961): 318–56.

10 J. Monod, *Chance and Necessity: An Essay on the Natural Philosophy of Modern Biology* (New York: Vintage, 1972).

specifically in favor of legalizing the birth control pill and abortion. Monod was an outspoken advocate for racial equality, and introduced Martin Luther King Jr. at the March 1966 meeting in Paris of the "Movement for the Peace." It was just one year before the great human rights leader was murdered. Monod took part in French student protests in the late 1960s, joining sit-ins at the Sorbonne and condemning the police brutality that followed. He was strongly critical of the stifling authoritarian mode of higher education in France.

Jacques Monod delved into philosophy, being highly influenced by his friend Albert Camus. He wrote about his conviction that all of the complexity of diverse forms on Earth derived from accidentally lucky chance molecular collisions and mutations.[11] Jacques Monod died of leukemia in 1976 at the age of sixty-six.

Lwoff's discovery of the two viral phases became the foundation for research into how some viruses, called oncogenic, can cause cancer. Lwoff's forty-seven-year tenure at the Pasteur Institute overlapped with a six-year stint as a professor of microbiology at the Sorbonne. Upon retirement from the Pasteur Institute in 1968, Lwoff directed the Cancer Research Institute at Villejuif, France, until 1972. André Lwoff died in 1994 at the age of ninety-two.

François Jacob's understanding of the research and discovery process ended up very different from his starting viewpoint. "The process of science does not consist in explaining the unknown by the known," he said. "It aims on the contrary, to give an account of what is observed by the properties of what is imagined—to explain the visible by the invisible. And it is through an appeal to new hidden structures with hypothetical properties that science proceeds."[12] In short, imagination precedes discovery. François Jacob, who had been the youngest of the 1965 prizewinners, died in 2013 at the age of ninety-two.

---

11 J. Monod, *Chance and Necessity: An Essay on the Natural Philosophy of Modern Biology* (New York: Vintage, 1972).

12 F. Jacob, *Of Flies, Mice and Men: On The Revolution in Modern Biology* (Cambridge, MA: Harvard University Press, 2001).

16

The Cancer Detectives

The 1966 Nobel Prize in Medicine was shared by Peyton Rous and Charles Huggins for their separate discoveries about cancer.

In 1966, Peyton Rous was eighty-seven years old when he stood on a chair raising a glass, cheering "Skoal!" late into a night of celebration with a thousand Swedish medical students gathered in his honor. Indeed, generations of medical students had studied the great discoveries of Peyton Rous when the prize was awarded, fifty-five years after his prizewinning research.

Peyton Rous was born in Baltimore in 1879, a descendant of Huguenots who had come to the New World to escape religious persecution in France. Rous earned a bachelor's degree at Johns Hopkins University. In his second year of medical school, he scraped his finger on a tuberculosis-ridden bone he was dissecting. He developed tuberculosis, a bacterial infection that primarily affects the lungs but can be spread to other parts of the body, as indeed it did in Rous, whose lymph nodes were infected.

There was no drug treatment for TB, for it was still some forty-five years from the development of the first antibiotic for it. Some doctors recommended "purging," with techniques such as bloodletting or dosing with herbs to force bowel movements or vomiting. Luckily, Rous took the other popular treatment, which was simply to get away from the city, eat well, and engage in outdoor exercise eight to ten hours per day. He took a year's leave from medical school and moved to Texas to work as a herdsman and ranch hand, even sleeping on the ground with the other cowboys. He made a full recovery and returned to Johns Hopkins to complete his medical degree.

Infectious Disease from Corpses

The very recently dead can harbor actively infectious particles that can be transmitted in the same way as in the living—through exposure to infected tissues, blood, weeping wounds, airborne particles, and body fluids such as saliva, urine, and stool. For the medical examiner doing an autopsy or even for an embalmer at a funeral home, the greatest risk is upon making the first cuts to enter the body, where contents may be under pressure and expel a burst of air or fluid. Less commonly, infectious agents are spread by infestations, such as by lice that carry the typhus organism or fleas carrying plague bacteria.

Most disease-causing microorganisms cannot survive long in a dead body. Yet, the bodies of persons who die from typhoid or cholera should not be buried near water sources due to a short period when they can leak infectious particles. People handling corpses are at high risk for infections with the bacteria causing tuberculosis and with the group A streptococcus that causes a type of highly infectious meningitis. Hepatitis A virus can be caught from feces, while hepatitis B and C are transmitted through blood and body fluids, as is HIV virus. HIV can live for several days after body death.[1] Ebola and Marburg viruses are highly infectious, and outbreaks have been exacerbated by the traditional methods of body handling in preparation for burial. These viruses are thought

---

1   P. N. Hoffman and T. D. Healing, "The Infection Hazards of Human Cadavers," Guide to Infection Control in the Healthcare Setting, International Society for Infectious Diseases, isid.org/guide/, last updated February 2018.

to survive in dead bodies for up to a week.[2] There have also been reports of COVID infection from fresh corpses.

Perhaps the most menacing risk from a dead body is the transmission of prions, a poorly understood transmissible agent that might be a protein. These cause rapid neurological deterioration that is always fatal. The human form is called classic Creutzfeldt-Jakob disease (CJD), which can be familial or sporadic. Scrapie is a prion disease of sheep, and mad cow disease is a type of prion disease in livestock that feed on scrapie-infected sheep parts. Humans who consume the sick beef can get a variant of Creutzfeldt-Jakob disease. Prion particles have recently been detected in the skin of patients with CJD.[3]

After medical school, Peyton Rous was awarded a grant to study at the Rockefeller Institute for Medical Research, where he would stay for his entire career. Soon Rous took over the laboratory for cancer research. He had seen the 1908 report by two Danes, Ellerman and Bang, who demonstrated that leukemia in chickens could be passed to other chickens by injecting filtered blood from the cancerous birds.[4] This suggested leukemia was caused by a transmissible agent too small to be caught by the finest filters. The work garnered little attention in the medical world because at that time leukemias were not understood to be cancerous diseases. In fact, leukemia is a cancer in which the bone marrow makes abnormal and insufficient white blood cells.

Rous devised experiments to study the cause of chicken sarcoma, cancers that grow in the connective tissues, such as skin and bone. He extracted some tumor material, minced it up, and added salt water

---

2   J. B. Prescott, T. Bushmaker, R. J. Fischer, et al., "Postmortem Stability of Ebola Virus," *Emerging Infectious Diseases* 21, no. 5 (2015): 856–59.

3   A. Nihat and S. Mead, "Detection of Creutzfeldt-Jakob Disease Prions in Skin: Implications for Healthcare," *Genome Medicine* 10, no. 1 (2018): 22.

4   V. Ellerman and O. Bang, "Experimentelle Leukamie bei huehnern" [Experimental leukemia in chickens], *Zentralblatt für Bakteriologie, Mikrobiologie und Hygiene. Series A: Medical Microbiology, Infectious Diseases, Virology, Parasitology* 46 (1908): 595–609.

before passing it through a filter to remove cells and bacteria. The fluid that came through the filter was then injected into healthy chickens, and caused them to grow sarcomas.

At that time (1910–11), the transmissible agent in the fluid was called "a filterable particle," which Rous characterized as "a minute parasitic organism," although he didn't see it.[5] Despite some debate, this was generally supposed to be a virus. (It would not be until the 1930s that electron microscopes allowed a virus to be visualized.) Rous suspected that there was virus in the tumor liquid and that when he transplanted the infected fluid into the healthy chicken, the virus replicated inside cells to cause sarcoma. Moreover, when he studied the new tumor material, he again found the filtered liquid from it could in turn cause tumors in yet another crop of healthy chickens. Thus, Rous endorsed the theory of viral causation of tumors.

Rous's initial reports were met with disbelief and even derision, with the only possible explanation that his laboratory methods were shoddy and the samples must have been contaminated. Another argument was that if a cause was found for a tumor, then one must not be working with a cancer at all. Yet Rous undertook various versions of his basic experiments and concluded that viruses do initiate development and promote the growth of chicken sarcoma. He tried similar studies in mice but had no luck causing extracts of mouse tumors to pass cancer into other mice. After fifteen years of debate, other labs finally did replicate Rous's careful chicken experiments and reported they too found that virus extract from cancers did reliably cause tumors in previously healthy chickens. However, the phenomenon was relegated to the field of chicken diseases due to a widespread belief that they have nothing to do with sarcoma in mammals, much less human sarcoma or any other human cancer.

---

5   P. Rous, "A Transmissible Avian Neoplasm (Sarcoma of the Common Fowl)," *Journal of Experimental Medicine* 12 (1910) 696–705; P. Rous, "A Sarcoma of the Fowl Transmissible by an Agent Separable from the Tumor Cells," *Journal of Experimental Medicine* 13 (1911): 397–411.

Rous was discouraged enough to abandon cancer research for nearly two decades, turning to studies of the blood and liver. Blood typing had been discovered by an earlier Nobel Prize winner, but due to lack of a preservative and anticlotting agent, the only way to give a transfusion was directly from the vein of the donor into the vein of the recipient. In 1916, Rous, along with J. R. Turner, introduced a citrate-glucose solution that permitted storage of blood for up to four weeks after collection. This made it possible to have blood-transfusion depots, which were first established in 1917 near the front lines of WWI in Belgium—these became the precursor to modern blood banks.

Rous became interested in cancer research again in 1933, when his Rockefeller colleague Richard Shope made a discovery of the cause of warty growths on wild rabbits, which sometimes transformed into true cancers. Shope had demonstrated that an infectious virus was responsible for the papilloma tumors on rabbits. Paralleling Rous's sarcoma experiments, Shope was able to extract filterable particles (virus) from the rabbit tumors and inject them into healthy rabbits to cause cancerous growths.[6] Shope had finally found a mammalian model of cancer propagation by virus, prompting Rous to reenter the field of viral causation of cancer.

Meanwhile, the initial virus Rous had isolated from chicken sarcoma had continued to be propagated in labs and was extensively used in experiments in many different species. Eventually the so called Rous sarcoma virus (RSV) came to be used in cancer research labs the world over. However, even into the late 1950s there was doubt about the significance of Rous's initial 1911 RSV findings. At the 1958 Seventh International Congress against Cancer, the Rous sarcoma virus was still attributed by some to be laboratory artifact, and some said it had only served to mislead cancer researchers.[7]

---

6   R. E. Shope and E. W. Hurst, "Infectious Papillomatosis of Rabbits: With a Note on the Histopathology," *Journal of Experimental Medicine* 58, no. 5 (1933): 607–24.

7   J. S. Henderson, "A View from the Center of a World," in *A Notable Career in Finding Out: Peyton Rous (1879–1970)* (New York: Rockefeller University Press, 1977).

RSV is, in fact, so reliable in causing cancer in various cell cultures that it has served as the incubator in which to study molecular mechanisms in cancer cells. Two subsequent Nobel Prizes would be given for cancer discoveries that came out of studies using RSV. This research has yielded the genetic-mutation theory of cancer causation, which supposes that each cancer has its origin in the DNA of a single cell that has undergone a rare mutation to make that cell cancerous, setting the progeny of that cell on the path to form a cancer. This theory is supported by many observable facts: the ability to cause cancer-yielding mutations in the lab; the finding that cancer cells often harbor large numbers of mutant genes and chromosomal abnormalities; and the knowledge that many cancer-causing substances cause mutations.

The genetic-mutation theory gained dominance by the late 1950s, but Peyton Rous did not agree with it. In his 1966 Nobel acceptance speech, he said, "Numerous facts, when taken together, decisively exclude this supposition."[8] The theory has been challenged because of these equally compelling facts: about 50 percent of known cancer-causing chemicals do not cause genetic mutations; inflammation and trauma have been shown to cause cancers in the absence of any genetic mutations; mutations caused by cancer-causing chemicals are only rarely harmful; most of these mutations are adaptive, that is, they change gene expression in order to adapt to the toxic substance such that there is less damage to the body. These observations have given rise to a number of other theories that include many of the components of tumor formation proposed by Rous throughout his decades-long career in studying cancer. His various suggestions had in common the idea that it requires more than one event for cancer to arise. Perhaps a virus was in the body in a silent stage before erupting only when the host was most susceptible. Or chemical promoters might alter the environment of the virus to make it carcinogenic. Promoters might also be physical, such as a sunburn or a wound, both of which naturally

8   P. Rous, "The Challenge to Man of the Neoplastic Cell," December 13, 1966, NobelPrize.org, nobelprize.org/prizes/medicine/1966/rous/lecture/.

stimulate cell growth needed for repair. But when regulation of this growth goes wrong, a cancer develops. Rous identified several definite cancer initiators and cancer promoters. These ideas have given rise to discoveries by others of carcinogens causing alterations in cell-to-cell signaling that only secondarily result in changes in the genes.[9]

Rous did not like to use the term *discovery*, as he thought it was a self-important concept, instead explaining that he was "in the business of finding out."[10] He was a careful researcher who demanded detailed observations and data reporting from his students and from the constant of stream of outstanding researchers who populated his Rockefeller lab for months to years at a time. Rous was especially keen on precision communication of scientific fact, and he spent long hours writing his own papers and rewriting the papers of his juniors. However, unlike many in comparable leadership positions, Rous was extremely generous in allowing unknown junior researchers to be the first-named authors on publications—indeed, when they had done the hands on work, Rous's name often appeared only at the very end of the credits or not at all.[11]

Peyton Rous continued at the Rockefeller Institute until mandatory retirement at the age of sixty-five in 1945, but he conducted active cancer research as a member emeritus for the rest of his life. Rous was nominated for the Nobel Prize seventeen times from 1926 to 1951. Even his son-in-law Alan Hodgkin won a Nobel Prize before him, though he was not yet born when Rous made his initial discovery.

There were several factors contributing to how long it took for the work of Peyton Rous to be recognized with a Nobel Prize. His contemporaries in 1911 were so unwilling to consider the possibility of an unseen infectious particle causing cancer that they scoffed at his initial

---

9   P. Rous, "Surmise and Fact on the Nature of Cancer," *Nature* 183 (1959).

10 C. Huggins, "His Business Was Finding Out," in *A Notable Career in Finding Out: Peyton Rous (1879–1970)* (New York: Rockefeller University Press, 1977).

11 P. D. McMaster, "His Sparkling Versatility," *A Notable Career in Finding Out: Peyton Rous (1879–1970)* (New York: Rockefeller University Press, 1977).

findings, accusing him of shoddy laboratory methods. Then, a world war interrupted. When his findings in chickens were finally replicated by others, their importance to cancer in mammals was not recognized. It was twenty-one years after Rous's chicken studies that Shope found a virus in a mammalian tumor. Then, in the late 1930s, viruses had the potential to become more interesting when these "filterable parasitic agents" were first seen by electron microscope in Germany. Once again a world war intervened, with the effect of suppressing awareness of this remarkable new tool for many years.

The realization of the cancer potential of viruses had to wait for the discoveries of virus behavior and control within bacterial cells in the late 1940s. The findings of Nobel winners Lwoff, Monod, and Jacob finally supplied the rationale that Rous had only guessed at— there were suppressors and promoters at work, and viruses could be quiet and only at some later opportune moment become virulent. It wasn't until the 1950s that the RSV virus was shown to alter the shape of normal host cells it infected, and it was later still that the viral gene in RSV that triggers uncontrolled growth was found. Ultimately, it was discovered that parallel mechanisms take place in human cells that transform into cancer.

Peyton Rous died in 1970 after a brief illness with cancer. He was ninety years old.

Viruses Associated with Cancer[12]

Hepatitis B virus and hepatitis C virus cause liver infection that can sometimes lead to liver cancer.

Epstein-Barr Virus (EBV) is a common virus, with most people getting infected with it at some point in their lives. Several cancers are linked with EBV, including Burkitt's lymphoma, nasopharyngeal carcinoma (cancer of the upper throat), Hodgkin's lymphoma

---

12 J. T. Schiller and D. R. Lowy, "Virus infection and Human Cancer: An Overview," *Recent Results in Cancer Research* 193 (2014):1–10.

and non-Hodgkin's lymphoma, T-cell lymphomas, posttransplant lymphoproliferative disorder (too many white blood cells), and leiomyosarcoma (cancer in the soft tissue). Human papillomavirus (HPV) is a group of more than two hundred viruses, and at least a dozen of them have been associated with cancer of the cervix, vulva, vagina, penis, anus, tonsils, or tongue. Kaposi's sarcoma–associated herpesvirus can cause Kaposi's sarcoma, a cancer of the blood vessels, as well as two types of lymphoma. This relationship is usually only found in immune-compromised patients, such as persons who have undergone an organ transplant, receive chemotherapy, or have AIDS.

Human T-cell lymphotropic virus type 1 (HTLV-1) infects T cells, which are a type of white blood cell. It can cause leukemia and lymphoma in about 2 to 5 percent of infected persons.

Human immunodeficiency virus type 1 (HIV-1, or HIV) causes a weakened immune system and confers a greater chance of viral-associated cancers such as Kaposi's sarcoma, non-Hodgkin's lymphoma, and cervical cancer. Merkel cell polyomavirus is a common virus that infects the skin, but in rare cases it causes skin cancer.

Simian virus 40 (SV40) was originally an infection of rhesus macaques imported from India for use in research labs, and it can cause cancer in monkeys. For the early versions of the human polio vaccine, poliovirus was grown on SV40-infected macaque kidney tissue, resulting in contamination of 10 to 30 percent of polio vaccines distributed from 1955 to 1963. Individuals born long after the use of contaminated vaccines have displayed evidence of SV40 infection. This demonstrates that the SV40 that first infected humans through the polio vaccine has since become human-to-human transmissible. SV40 has the potential to cause cancer in humans, with the finding of SV40 within certain cancer tumors ranging from 0 to 56%.[13]

---

13 J. S. Butel, "Patterns of Polyomavirus SV40 Infections and Associated Cancers in Humans: A Model," *Current Opinion in Virology*, 2 no. 4 (2012): 508–14.

Avian leucosis virus (ALV) is the current terminology for the types of viruses discovered by Ellerman and Bang in 1908 causing leukemia in birds and then by Rous in 1911 causing sarcoma in chickens. Viruses propagated on cells from chicken embryos have long been used in vaccine development. ALV contaminated the yellow fever vaccine in the 1960s, which was addressed by ultrafiltration to remove ALV. Some manufacturers have changed to using certified ALV-free embryo cells, but as of the twenty-first century, some measles, mumps, rubella vaccines and influenza vaccines are still routinely produced with ALV contamination. There is no evidence for the ALV virus causing cancer in humans, but due to the way the virus interacts with cancer-controlling human genes, the potential to promote cancer remains a possibility.[14]

Charles Brenton Huggins was awarded the other half of the 1966 Nobel Prize for his discoveries concerning the influence of hormones and the use of hormonal treatment for some cancers. Huggins was born in Halifax, Canada, in 1901. After his college years in Canada, Huggins attended Harvard Medical School and then studied surgery at the University of Michigan. In 1927, he joined the faculty of the newly formed medical school at the University of Chicago, where he spent the rest of his career.

Huggins specialized in surgery of the male genitourinary system, so he saw many prostate cancers. Before his discoveries in 1940, it was widely assumed that cancers were autonomous and self-perpetuating. It was thought that cancers were on an automatic cell-dividing rampage that was independent of the host conditions—the biochemical

14 M. A. Raines, N. J. Maihle, C. Moscovici, L. Crittenden, and H. J. Kung, "Mechanism of c-erbB Transduction: Newly Released Transducing Viruses Retain Poly(A) Tracts of erbB Transcripts and Encode C-terminally Intact erbB Proteins," *Journal of Virology* 62, no 7 (1988): 2437–43.

physiology of the rest of the body. (Rous's theories were an exception to the mainstream, as Rous had suggested that host factors contributed to the viral causation of cancers.)

Huggins and his two medical student assistants made careful measurements of the hormones in the body and discovered that testosterone was necessary for normal prostate cell growth and function. But they saw a different effect of testosterone on prostate cancer cells: testosterone promoted aggressive cell growth in prostate tumors.

Huggins experimented on dogs because they are the only species besides man that gets prostate cancer. He deprived their bodies of testosterone by castration, causing their prostate tumors to shrink and sometimes disappear. Hormone measurements also revealed that estrogen, the hormone that predominates in females, exists in small amounts in males. Huggins found that giving extra estrogen to dogs diminished the effect of normal testosterone, and it was an effective alternate way to shrink their prostate cancers. Huggins found the same results in humans. The effect was especially remarkable on prostate cancers that had metastasized to the bones, where the tumors caused pain and threatened spontaneous fractures. Sometimes the treatment would bring relief of bone pain within hours and, weeks later, the regression of bone lesions seen on X-ray.[15] This became the first cancer chemotherapy. Today advanced prostate cancers are still treated hormonally. Testosterone and related compounds are diminished by surgical castration, or more commonly by drugs that interfere with testosterone production.

Huggins also did extensive work on breast cancers, first experimenting with mammary cancers in lab rats then conducting tests on women with breast tumors. He built on the theory of Lacassagne, who

---

15 C. Huggins and C. V. Hodges, "Studies on Prostatic Cancer. I. The Effect of Castration, of Estrogen and of Androgen Injection on Serum Phosphatases in Metastatic Carcinoma of the Prostate," *Journal of Urology* 167, no. 2 (1941): 948–51; C. Huggins, R. E. stevens Jr., C V. Hodges, "Studies on Prostatic Cancer II. The Effects of Castration on Advanced Carcinoma of the Prostate Gland," *Archives of Surgery* 43, no. 2 (1941): 209–23.

in 1932 had demonstrated that estrogens can promote some mammary tumors in mice.[16] This led to removal of the ovaries—the source of most female estrogen—as a treatment for breast cancer. Huggins repeated the rodent experiments and was able to demonstrate that depriving or blocking estrogen caused death to the cancer cells while the normal cells shrunk but survived. Then he experimented with giving excess hormones—in this case estrogen and progesterone—which likewise caused cancer cells to die but caused exuberant growth in normal cells, just as the breasts enlarge when hormones peak naturally during pregnancy.[17] He reasoned that the excess of hormones acted as "hormone-interference," where normal tissues took up the hormones preferentially, leaving the tumor cells to die.

Huggins urged his junior colleague Elwood Jensen to conduct experiments to determine which tumors would be hormone sensitive. Jensen went on to discover estrogen receptors—the chemical configuration on a cell which, like a lock and key, accepts estrogen.[18] Today women with breast cancer have tumors tested to see if the cell surfaces have chemical receptors for hormones. This helps to customize treatment and has revolutionized breast cancer therapies. Hormone-sensitive breast cancer is currently treated in many ways. Women who are premenopausal with functioning ovaries are offered ovarian-suppressing drugs or surgical removal or irradiation of the ovaries. Postmenopausal women who have tumors bearing estrogen receptors are offered drugs that block the activity of an enzyme that makes estrogen. The most commonly prescribed antiestrogen drugs bind to estrogen receptors in order to block estrogen from getting into the cell, while some other drugs cause destruction of the receptor.

---

16 A. Lacassagne, "The Relation between Hormones and Cancer," *Canadian Medical Association Journal* 37, no. 2 (1937): 112–17.

17 R. L. Landau, E. N. Ehrlich, and C. Huggins, "Estradiol Benzoate and Progesterone in Advanced Human-Breast Cancer," *JAMA* 182, no. 6 (1962): 632–36.

18 E. V. Jensen, E. R. De Sombre, "Oestrogen-Receptor Interaction in Target Tissues," *Biochemical Journal* 115, no. 5 (1969): 28P–29P.

In addition to prostate cancer and some breast cancers, other sex-hormone-dependent tumors are cancers of the testicles, ovaries, and uterus. Thyroid cancer and osteosarcoma (cancer of the bones) can be dependent on hormones other than sex hormones.

Huggins concluded, "The net increment of mass of a cancer is a function of the interaction of the tumor and its soil. Self-control of cancers results from a highly advantageous competition of host with his tumor."[19]

Huggins was nominated for the Nobel Prize three times in 1950 and 1951. It is a mystery why the Nobel Committee waited until 1966, fifteen years after his last nomination, to recognize his earlier work, especially since his discoveries were immediately demonstrated to be applicable in treating human cancer.

Charles Huggins died in 1997 at the age of ninety-six.

---

19 C. Huggins, "Endocrine-Induced Regression of Cancers," December 13, 1966, NobelPrize.org, nobelprize.org/prizes/medicine/1966/huggins/lecture/.

17

The Visionaries

The 1967 prize was shared by Ragnar Granit, Keffer Hartline, and George Wald for their independent discoveries regarding vision.

Ragnar Granit was a Finnish Swede born in 1900 in Finland to Swedish-speaking parents. This was not unusual, as Finland was part of Sweden for some seven hundred years before Russia took over much of the country in the early nineteenth century. Finland declared independence in 1917, which delighted German allies but did not go over well with the Russians. As a teenager, Granit took part in Finland's civil war on the side of the Whites, who were allied with Germany and fought the Russian-backed Reds. The Whites prevailed, and Finland stabilized as an independent nation in 1918.

After completing medical school in Helsinki, Granit studied twice at Oxford with Charles Sherrington, a Nobel laureate and the most renowned brain and nerve researcher of the day. In between Oxford stints, Granit started research on the nerve mechanisms of the eye at the University of Pennsylvania. In 1935, Granit became a professor at Helsinki, where he continued his research on eyesight. His career was interrupted in late 1939, when Granit was temporarily assigned to be a physician to troops while the Russians invaded again in the short Winter War. During a brief hiatus in hostilities in 1940, Granit accepted an offer to join the Royal Caroline Institute of Stockholm. He got out just in time. The so-called Continuation War, fought between the Russians and an alliance of Finns with the German army, would persist until 1945. Granit became a Swedish citizen in 1941, and he remained on staff at the Royal Caroline for the rest of his career.

Granit's prizewinning research was on the ways in which the eyes and brain interact with light. The backs of the eyes have special light-sensitive cells: rods, which help vision in low light, and cones, which are sensitive to color. The eye is least sensitive to red, which explains why red objects appear to dull and go gray-black when in lower light. Blue cone cells are the most sensitive to light, which is why a field of blue flowers will still look vivid even as the sun sets. The back of the eye is an extension of the brain. When light plays on the rods and cones, it activates chemical and electrical signals that go through a complicated relay system to areas of the brain.

Through painstaking electrical recording, Granit discovered that some nerve fibers in the brain are sensitive to the entire spectrum of light. These "dominators" detect the intensity of light but not the color. Other nerve fibers are sensitive to narrow color-specific bands of light. These "modulators" detect color. The interplay between dominators and modulators not only determines color perception but also contributes to contrast, helping show detail in objects. The most surprising of Granit's discoveries was that light not only stimulates nerve impulses but can inhibit impulses along the optic nerve that connects with the brain.[1] This is a similar design to other sense organs, including the skin, nose, tongue, and ear (as described in the chapter on the 1961 Nobel Prize).

Granit eventually found the work of recording light sensitivities to be too monotonous, so he capped his achievements by writing textbooks on the subject and then left the field of visual research.[2] By the

---

1   R. A. Granit, *Sensory Mechanisms of the Retina* (London: Oxford University Press, 1947).

2   R. A. Granit, "Neurophysiology of the Retina," in H. Davson, ed., *The Eye* (New York: Academic Press, 1962), 534–796.

time Granit gave his Nobel lecture in 1966, he had been involved for many years in the study of the regulation and control of muscle action.

In 1977, Granit wrote a comprehensive collection of essays, *The Purposive Brain*, on how the central nervous system interacts with the environment. While it is detailed enough to interest a graduate student of neurophysiology, it is also approachable for anyone with a basic science education—provided they are armed with a good dictionary. Granit described that his vast experience of penetrating the brain with electrodes led him to conclude that this field of study does not explain consciousness. He quotes Polanyi's example of the hierarchical model of language development as an analogy to the progressive layers of physical, electrical, and biochemical controls found in the human nervous system: "Voice production leaves largely open the combination of sounds into words, which is controlled by a vocabulary. Next, a vocabulary leaves largely open the combination of words to form sentences, which is controlled by grammar, and so on. . . . You cannot derive a vocabulary from phonetics; you cannot derive grammar from a vocabulary; a correct use of grammar does not account for good style and a good style does not supply the content of a piece of prose."[3] Simply put, the laws governing lower levels of operation do not control or predict higher levels of operation. The discoveries of various physical, biochemical, and electrical phenomena in the nervous system do not account for what happens at a higher level. Granit said, "Conscious man makes use of neurophysiological mechanisms without being governed by them." He found that "the properties of conscious awareness cannot be explained by any of the numerous physiological properties within reach of our technical prowess," and concluded, "it is a futile occupation to hunt for structural parallelism of likenesses between the

---

3   M. Polanyi, "Life's Irreducible Structure: Live Mechanisms and Information in DNA Are Boundary Conditions with a Sequence of Boundaries above Them," *Science* 160, no. 3834 (1968): 1308–12.

physiological processes and conscious awareness when all we can do is to establish correlations."[4]

Ragnar Granit died in 1991 at the age of ninety.

———————

Haldan Keffer Hartline was born in Pennsylvania in 1903. He had an interest in visual systems early on, but after medical school, he realized he needed more training in mathematics and physics to further his experiments on the eye and brain. He thereby became a major contributor in the new field of medical physics.

Hartline and his team at the University of Pennsylvania were the first to place electrodes in fibers of the optic nerve, using a horseshoe crab as their subject. These strange creatures have eyes protruding from either side and in the middle of their carapace, each connected to the brain by a large, long optic nerve. In Hartline's experiments, the optic nerve of a lateral eye was frayed into thin bundles, which were then split until just one active fiber remained. The single nerve fiber was measured as the eye was stimulated with light. This simple experiment demonstrated that nerve impulses relay visual information to the brain.

Subsequent experiments revealed a much more complex system than simple "on" impulses excited by light. Hartline found that individual nerve fibers do not all react the same way. Some respond only for a short time when light first hits the eye, again anytime there is a fluctuation of the intensity of light, and again when the light is turned off. These nerve fibers are otherwise inactive in between light fluctuations. Other specialized nerve fibers are stimulated by steady light. Other fibers give no response during initial illumination, firing a vigorous and prolonged train of impulses only when light is dimmed. Humans experience this latter phenomenon as the gradual adaptation to a dark room when the lights suddenly go out.

———————

4   R. Granit, *The Purposive Brain* (Cambridge, MA: MIT Press, 1977).

Hartline found that the light receptors are highly adaptable to change, with nerve impulses rapidly transmitting when a light first appears, followed shortly by a slower rate of impulse transmission as the light continues to shine. Thus, we wince when first stepping outdoors into the sun, followed by rapid adaptation to the bright light so we are soon no longer avoiding it. Hartline found that adaptation occurs continuously: when the light is very intense, very dim, and on all gradients in between.

Hartline discovered that the light-sensitive cells at the back of the eye are connected by an elaborate but purposeful network of nerve fibers; these serve to inhibit nearby transmissions, dampening the effect of light excitation on neighboring cells. The influences are mutual: "Each receptor, being a neighbor of its neighbors, inhibits and is inhibited by those neighbors."[5] But the strength of the inhibition is not equal: Receptors that are strongly excited by bright light exert a stronger inhibition on neighboring receptors in more dimly lit regions, while the dimly lit region exerts less of an inhibition on the receptors receiving the brighter light. The resultant visual effect is to have a greater contrast between light and shadow, sharpening the edges of shapes and overall giving us a crisper impression of the world.

Where does mathematics come into all this? The experiments were rather simple, with light shining on an eye and electrodes placed in the retina and along a network of nerve fibers leading to the optic nerve, producing a series of dots and dashes and curves on an oscilloscope. But it required mathematics for Granit to translate what he observed into language that communicated about vision, with equations such as this:[6]

$$r_p = e_p - \sum_{j=1}^{n} K_{p,j}(r_j - r^0_{p,j}) \qquad p = 1, 2, \cdots n$$

5   H. K. Hartline, "Visual Receptors and Retinal Interaction," December 12, 1967, NobelPrize.org, nobelprize.org/uploads/2018/06/hartline-lecture.pdf.

6   H. K. Hartline and F. Ratliff, "Spatial Summation of Inhibitory Influences in the Eye of Limulus, and the Mutual Interaction of Receptor Units," *Journal of General Physiology* 41, no. 5 (1958): 1049–66.

In his Nobel banquet speech, Hartline noted the emphasis in Alfred Nobel's last will and testament on the tangible benefits of scientific discoveries. Though Hartline admitted that his discoveries had no obvious applications, he acknowledged the wisdom of the Nobel Committee in recognizing the importance of basic research simply in the interest of adding to the understanding of our universe and ourselves.[7] He said, "I am no linguist and know only two Swedish phrases—the first is but a single word, internationally understood, and used wherever glasses are raised; the second I will use now: Tack så mycket [Thank you very much]."

Keffer Hartline died in 1983 at the age of seventy-nine.

George Wald was born in New York City in 1906. After getting a degree in science, he did his graduate work in the field of zoology. He became interested in the biochemistry of the eye and particularly intrigued by the biology behind night blindness, which was already attributed to a lack of vitamin A. His first wife, Frances, was instrumental in his studies, procuring thousands of retinas from cattle, pigs, and sheep for his experiments.

In the early 1930s, Wald obtained a travel grant to study in the German lab of Nobel laureate in medicine Otto Warburg. There, Wald investigated the composition of rhodopsin, a light-sensitive pigment within the cells of the retina. He discovered that vitamin A was an important component of rhodopsin in the retina.[8] Wald continued these investigations in Switzerland under Paul Karrer, who had recently resolved the structure of vitamin A. (Karrer would receive a Chemistry Nobel Prize in 1937.) Then Wald spent a year in the lab of Otto

---

7   K. Hartline, Banquet Speech, December 10, 1967, NobelPrize.org, nobelprize.org/prizes/medicine/1967/hartline/speech/.

8   G. Wald, "Vitamin A in Eye Tissues," *Journal of General Physiology* 18 (1935): 905–15..

Meyerhof, another Nobel laureate in medicine. Wald described the time he spent in the laboratories of these three Nobel laureates as opening up a new life for him—a life with molecules. As he described it, "From then on it has been a constant going back and forth between organisms and their molecules—extracting the molecules from the organisms, to find what they are and how they behave, returning to the organisms to find in their responses and behavior the greatly amplified expression of those molecules."[9]

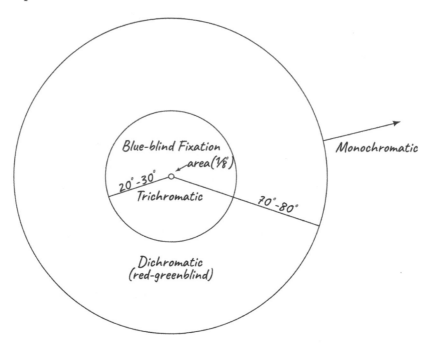

Wald's research contributed to the discovery that the human retina senses color unevenly. There is no perception of blue at the fovea (the central focusing spot), as this region lacks blue-sensitive cone cells. Over a spread of twenty to thirty degrees out from the center, the retina has all three color-sensing cones: red, green, and blue. From there, it transitions to red-blind and green-blind, until about seventy to eighty degrees out, at which point vision is monochromatic.

---

9   G. Wald, "The Molecular Basis of Visual Excitation," December 12, 1967, NobelPrize.org, nobelprize.org/uploads/2018/06/wald-lecture.pdf.

Wald returned to America and eventually joined the staff at Harvard. Over three decades, Wald conducted experiments to determine how light causes rhodopsin to change form and give rise to signals that culminate in visual impressions. Much of this research was conducted with Ruth Hubbard, who in 1958 would become his second wife. Hubbard had attended Radcliffe College in 1940, a sister institution to Harvard, where undergraduate women were not yet accepted. She first joined the laboratory of George Wald during the war, and she rejoined it as a research fellow after earning her PhD in biology at Harvard (the graduate school had accepted women since 1920). Her thesis involved vitamin-A-related compounds in the retina. She eventually became faculty in the Department of Biology at Harvard, but like most women in academic science at that time, she had limited opportunities for advancement. In the same year that Wald won the Nobel Prize, both Wald and Hubbard were awarded the Paul Karrer Gold Medal for their work on eye pigments.

Ruth Hubbard became politically active in the mid 1960s. She joined a group that petitioned Harvard to reevaluate the job statuses of its female faculty, and in 1973, she became the first woman to be offered a tenured Harvard professorship in the Biology Department. She wrote and spoke widely on women and science, not just about equal opportunity but also challenging the conventional direction of male-dominated scientific inquiry. In referring to persons other than white males, she said, "Though we have been described by scientists, by and large we have not been the describers and definers of scientific reality. We have not formulated the questions scientists ask, nor have we answered them. This undoubtedly has affected the content of science, but it has also affected the social context and the ambience in which science is done."[10]

Wald and Hubbard had a son, Elijah, who described how the recognition and notoriety afforded by the Nobel Prize effected a dramatic

---

10 F. B. Fiske, "Scholars Face a Challenge by Feminists," *New York Times*, November 23, 1981.

shift in his father's focus, largely away from scientific research and more toward political activism. On March 4, 1969, George Wald gave a speech during a strike by scientists at Massachusetts Instititute of Technology (MIT), the nation's largest academic defense contractor at that time. He spoke about American war crimes, including the Vietnam War; the illogic of a peacetime military draft; the militarization of America as part of a military-industrial-labor-union complex; and the absurdity of nuclear war on a small planet.[11] The speech, "A Generation In Search of a Future," was published in full by the *Boston Globe*, which subsequently distributed eighty-seven thousand copies in requests for reprints before they stopped due to cost. The speech was also eventually carried in the *New Yorker*, the *Progressive*, the *New York Post*, the *Washington Post*, the *Philadelphia Bulletin*, the *Chicago Daily News*, the *Buffalo Evening News*, the *San Francisco Chronicle*, and other papers. It is estimated that millions of people read it. One part of the speech that was omitted (at Wald's request) in subsequent reproductions was Wald's offer to assist draft dodgers in any way he could. This statement potentially implicated him in what was then a felony in the United States—one for which others were being actively arrested. The unedited version was still available online as of February 2021.[12]

> I think the Vietnam war is the most shameful episode in the whole of American history. The concept of war crimes is an American invention. We've committed many war crimes in Vietnam—but I'll tell you something interesting about that. We were committing war crimes in World War II, before the Nuremberg trials were held at which the principle of war crimes was stated. The saturation bombing of German cities was a war crime. Dropping those atomic

---

11 G. Wald, *A Generation in Search of a Future* (Stoughton, MA: Press of Nightowl, 1969).

12 G. Wald, "George Wald–A Generation in Search of a Future," March 4, 1969, video, posted May 17, 2020, youtube.com/watch?v=5zDzjyDmqns&feature=youtu.be.

bombs on Hiroshima and Nagasaki was a war crime. If we had lost the war, it might have been our leaders who had to answer for such actions. I've gone through all that history lately, and I find that there's a gimmick in it. It isn't written out, but I think we established it by precedent. That gimmick is that if one can allege that one is repelling or retaliating for an aggression, after that everything goes.

And, you see, we are living in a world in which all wars are wars of defense. All War Departments are now Defense Departments. This is all part of the doubletalk of our time. The aggressor is always on the other side. I suppose this is why our ex-Secretary of State Dean Rusk went to such pains to insist, as he still insists, that in Vietnam we are repelling an aggression. And if that's what we are doing—so runs the doctrine—everything goes. If the concept of war crimes is ever to mean anything, they will have to be defined as categories of acts, regardless of alleged provocation. But that isn't so now.

I think we've lost that war, as a lot of other people think, too. The Vietnamese have a secret weapon. It's their willingness to die beyond our willingness to kill. In effect, they've been saying, You can kill us, but you'll have to kill a lot of us; you may have to kill all of us. And, thank heaven, we are not yet ready to do that.

Yet we have come a long way toward it—far enough to sicken many Americans, far enough to sicken even our fighting men. Far enough so that our national symbols have gone sour. How many of you can sing about "the rockets' red glare, the bombs bursting in air" without thinking, Those are our bombs and our rockets, bursting over South Vietnamese villages? When those words were written, we were a people struggling for freedom against oppression. Now we are supporting open or thinly disguised military dictatorships all over the world, helping them to control and repress peoples struggling for their freedom.[13]

In 1980, Wald was part of the delegation to Iran during the Iran hostage crisis. In 1986, Wald was part of a small group of Nobel laureates who met with Mikhail Gorbachev in Moscow to discuss environmental concerns. Wald took the opportunity to question Gorbachev about the exile of Nobel Peace Prize–winner Andrei Sakharov and his wife Yelena Bonner. Wald reported that Gorbachev said he knew nothing about it, but Bonner and Sakharov were released shortly thereafter.

George Wald on Nuclear War

"We have to get rid of those nuclear weapons. There is nothing worth having that can he obtained by nuclear war—nothing material or ideological—no tradition that it can defend. It is utterly self-defeating. Those atomic bombs represent an unusable weapon. The only use for an atomic bomb is to keep somebody else from using one. It can give us no protection—only the doubtful satisfaction of retaliation. Nuclear weapons offer us nothing but a balance of terror, and a balance of terror is still terror. We have to get rid of those atomic weapons, here and everywhere. We cannot live with them."[14]

Neurophysiologists, who probe the depths of the brain, are more prone than other medical researchers to philosophize about the animating principle of life. George Wald had a lot to say on the subject, and he concluded that consciousness has no location, cannot be measured, and is independent of space and time. "Perhaps it was in part because I am a biologist that the idea at first seemed so strange to me. Biologists tend to be embarrassed by consciousness. As it is an attribute

---

13 G. Wald, *A Generation in Search of a Future* (Stoughton, MA: Press of Nightowl, 1969).

14 G. Wald, *A Generation in Search of a Future* (Stoughton, MA: Press of Nightowl, 1969).

of some living organisms, they feel that they should know about it, and should indeed be in position to straighten out physicists about it, whereas exactly the opposite is true." Wald expanded on this notion:

> Though consciousness is the essential condition for all science, science cannot deal with it. That is not because it is an unassimilable element within science, but just the opposite: science is a highly digestible element within consciousness, which includes science as a limited territory within the much wider reality of whose existence we are conscious. Consciousness itself lies outside the parameters of space and time that would make it accessible to science, and that realization carries an enormous consequence: consciousness cannot be located. But more: it has no location.[15]

George Wald died in 1997 at the age of eighty-nine. His son Elijah maintains a web page on him, including the full text of an entertaining essay called "On the Origin of Death," in which Wald goes through the biological evolution of body death.[16]

---

15 G. Wald, "Life and Mind in The Universe," *International Journal of Quantum Chemistry* 26 (1984): 1–15, full text available at elijahwald.com/lifeandmind.html.

16 G. Wald, "The Origin Of Death," elijahwald.com/origin.html.

18

Making Protein

Marshall Nirenberg, Har Gobind Khorana, and Robert Holley won the 1968 Nobel Prize for each of their key discoveries in interpreting the genetic code. Working independently from one another, these three led laboratory experiments to discover just how DNA codes for the production of amino acids, from which proteins could be formed and the whole body manufactured.

By the time of the work of Marshall Nirenberg, enzymes that facilitate the biochemical reactions to make amino acid had been discovered. The component molecules of DNA were known. The structure of DNA had been determined. Many had postulated that a related substance, RNA, was a sort of intermediary messenger molecule interacting with DNA and the process of making of amino acids. But no one had put it all together yet.

Nirenberg's work showed that messenger RNA lines up with an untwisted portion of DNA and transcribes a sequence of its code. Khorana showed how RNA makes amino acids to build proteins. Holley provided the connection between the two by finding the structure of transfer RNA (tRNA), which brings or transfers amino acids to the cellular structure where they are assembled into proteins.

"Marshall Warren Nirenberg is the most famous man you have never heard of." So opens his biography written by a former laboratory colleague.[1] Nirenberg was born in New York City to a shirt manufacturer. He lived in Brooklyn until the age of ten, when he was made an invalid for a year because of rheumatic fever, a complication of infection with streptococcus bacteria in the pre-penicillin age. After a brief stint in Mount Vernon, New York, the family relocated to operate a dairy farm in rural Orlando, Florida.

Nirenberg attended the University of Florida in Gainesville, which in 1945 was not exactly a breeding ground for Nobel laureates. He graduated with a grade point average of just 2.32 out of 4, but was nevertheless admitted to the master's program at the university. His thesis involved the examination of 10,800 caddis flies, most of which he personally captured. He completed the zoology program in two years with a near-perfect academic record. Nirenberg was accepted as a PhD candidate at the University of Michigan at Ann Arbor, where he joined a relatively weak Biochemistry Department that was probably accepting candidates who could not get in at more prestigious institutions. Nirenberg did not make progress on his assigned project of nutritional biochemistry, and he was a constant disappointment to his graduate adviser. Finally, Nirenberg switched gears and devised experiments to study the way cancer cells metabolize sugars, which merited publication and earned his degree.

His mentor urged Nirenberg to apply for a fellowship at the National Institutes of Health (NIH) in order to improve his skills in biochemistry. At the NIH, Nirenberg was assigned to work with an eminent researcher on the biology of enzymes, but he became much more interested in the problem of the genetic code: What were the operations that resulted in instructing for every form and function of the human body? His background was in insect zoology and then in cancer-cell biochemistry, with no formal training in genetics. A friend warned it would be professional suicide to enter the field, but Nirenberg ignored him.

1   F. H. Portugal. *The Least Likely Man: Marshall Nirenberg and the Discovery of the Genetic Code* (Cambridge, MA: MIT Press, 2015).

Cracking the genetic code was the hottest challenge in biochemistry, engaging top researchers at the Pasteur Institute, Cambridge, Caltech, New York University, and Harvard. Whoever figured it out would be a shoo-in for a Nobel Prize; therefore, the most aggressive researchers in the business were very careful not to share information on their advances too freely, while at the same time, they tried hard to glean what competitors might be up to. Secrecy had to be balanced by opportune timing of the announcement of new discoveries, lest someone else beat you to it.

There was even a secret society to swap theories and limited information. Members of the secret RNA Tie Club included Crick and Watson, four more Nobel winners, and fourteen others. They wore ties embroidered with the helical molecule. Some of the Tie Club boys were strictly theorists, not getting their hands dirty with experimentation. Francis Crick was typical of this ilk—he'd helped to discern the structure of DNA via calculations based on the actual laboratory work of others without ever doing a single experiment on it. In contrast, Nirenberg, not a club member, steadfastly worked through experiments to get to some answers. Initially he had to conduct his decoding experiments on the side, unbeknownst to his seniors at the NIH, all the while keeping a personal journal that documented his methodical approach to cracking the code. Typical journal entries find Nirenberg ordering himself to buckle down, work longer hours, stop cogitating, and just do the work, whatever it takes. After two years he could finally research in the open when he was assigned his own lab. On the other hand, Crick enjoyed captivating his fellow club members with his various theories but could not confront the laboratory work it would take to make discoveries. In a private note circulated to RNA Tie Club members, Crick said, "In the comparative isolation of

Cambridge I must confess that there are times when I have no stomach for decoding."[2]

Nirenberg knew nothing of the secret club. He was a definite outsider, an unknown, not even at an academic institution. In 1961, he and his postdoctoral fellow J. H. Matthaei succeeded in demonstrating the function of messenger RNA (mRNA).[3] Shortly thereafter, they showed how mRNA utilizes a three-component code that results in the production of an amino acid. In essence, this was the first "word" decoded from genes.[4]

This discovery shot Nirenberg into the worldwide limelight, and he became acutely aware of sharp competition to find the codes for all of the amino acids. More than once announcements of his latest experimental results would get hijacked by other contenders—for example, he would be at a symposium winding up a presentation of recent findings when a member of the audience would feign to pose a question only to use the opportunity to turn it into a forum for their own recent discoveries. It took five years and a team of some twenty researchers in his lab—including many NIH scientists who put aside their own research to help the efforts—so as not to be outrun by others such as Severo Ochoa's team at New York University. In 1966, Nirenberg announced his crowning achievement: the codes for all basic human amino acids.

Taking the mystery out of it all, Nirenberg described messenger RNA as simply a robot that only followed the commands of DNA to

---

2   F. R. C. Crick, "On Degenerate Templates and the Adaptor Hypothesis: A Note for the RNA Tie Club" (privately circulated), January 1955, Francis Crick (1916–2004): Archives, PP/CRI/H/1/38, Wellcome Collection, wellcomecollection.org/works/j7qpsmdm.

3   M. W. Nirenberg and J. H. Matthaei, "The Dependence of Cell-Free Protein Synthesis in E. coli upon Naturally Occurring or Synthetic Polyribonucleotides," *Proceedings of the National Academy of Sciences of the United States of America*, 47 (1961): 1588–1602.

4   H. J. Matthaei, O. W. Jones, R. G. Martin, and M. W. Nirenberg, "Characteristics and Composition of RNA Coding Units," *Proceedings of the National Academy of Sciences of the United States of America*, 48, no. 4 (1962): 666–77.

carry out genetic instructions. "Man now understands the language of the civilization, has written quite elementary messages in the form that robots understand, and via such texts has communicated directly with the robot. The robots read and faithfully carry out the instructions."[5]

The hardworking Nirenberg enjoyed lifelong tenure at the NIH. As the genetic work was winding down, he moved into the field of neuroscience, ultimately researching how genes determine neurological anatomy and function. He worked with new methods of cloning neurological cancer-cell lines in order to provide live cells for study on the behavior of cancer. Some of his lab work contributed to the development of tests for neurological diseases. He made important contributions to understanding the brain biology of narcotic addiction. In another contrast to Crick, while Nirenberg's postprize activity was making real contributions in the field of neurobiology, Crick was unsuccessfully pondering theories of consciousness derived from the lab work others had done in the field of brain biochemistry.

Nirenberg leveraged his Nobel fame with episodic political activism, including protesting directly to the president of Brazil about the political purging of leading scientists, sending a plea to the president of the Soviet Union for the release of a hormone specialist who was being held as a political prisoner, and requesting the pope to press for an investigation into the disappearance of Argentinean physicists on his visit to that country. Nirenberg's opposition to the nuclear arms race and chemical warfare garnered the attention of presidents and politicians.

Marshall Nirenberg died at the age of eighty-two in 2010 after a brief illness with cancer.

---

5   M. W. Nirenberg, as quoted in "Translating the Code of Life and the Nobel Prize, 1962–1968," Profiles in Science, NIH National Library of Medicine, profiles.nlm.nih. gov/spotlight/jj/feature/codeoflife.

Another laureate with beginnings that would not predict a path to the Nobel Prize was Har Gobind Khorana, born in 1922 in a Punjab village of one hundred people. His father was the local tax clerk, and theirs was the only literate family in the region. Khorana related that his early education was outdoors under a tree, which modern studies show is highly therapeutic for youngsters.

The Benefits of Outdoor Learning

Studies have documented multiple benefits of outdoor classrooms, including academic, emotional, and physical advantages. Students in an outdoor classroom have been found to be more attentive and have a better recollection of the information that was studied.[6] Consistently exposing schoolchildren to nature has been shown to decrease stress and anxiety, and helps to elevate mood and emotion.[7] Exposure to bright sunlight is also healthy for vision and reduces nearsightedness in schoolchildren.[8]

At Punjab University, Khorana obtained his bachelor's and master's degrees in chemistry and biochemistry. A fellowship from the Indian government facilitated his PhD at the University of Liverpool in the UK, where he was introduced to the study of the building blocks of DNA and the proteins they code for. He pursued this line of research in Vancouver at the British Columbia Research Council, followed by a move to the University of Wisconsin. Khorana and his Wisconsin team worked out the mechanisms by which RNA codes for the synthesis of

---

6   E. Fägerstam and J. Blom, "Learning Biology and Mathematics Outdoors: Effects and Attitudes in a Swedish High School Context," *Journal of Adventure Education and Outdoor Learning* 13, no. 1 (2012): 56–75.

7   P. Bentsen and F. Søndergaard Jensen, "The Nature of Udeskole: Outdoor Learning Theory and Practice in Danish Schools," *Journal of Adventure Education and Outdoor Learning* 12, no. 3 (2012): 199–219.

8   Wu et al., "Myopia Prevention and Outdoor Light Intensity in a School-Based Cluster Randomized Trial," *Ophthalmology* 125 (2018): 1239–50.

proteins. They identified the specific biochemical rules for how sections of the DNA code that had been copied onto to RNA get translated into making amino acids and thus proteins.[9]

In 1970, Khorana moved to MIT, where he led a team to create the first totally man-made gene.[10] He retired in 2007 and was fondly remembered for his skills and kindness as a teacher. As of 2021, Khorana is the only Nobel Prize winner in medicine to be from India.

Har Gobind Khorana died in 2011 at the age of eighty-nine.

The discovery that connected the work of Nirenberg and Khorana was made by Robert William Holley, born in Urbana, Illinois, in 1922. He graduated from high school at sixteen, and four years later completed his undergraduate studies. Holley went to Cornell University for his PhD in organic chemistry and stayed on as faculty at Cornell. In the mid-1950s, he managed to take a sabbatical year to study at the California Institute of Technology. Toward the end of his work there, which was mainly on proteins, he completed a small experiment on amino acids. He was hooked—it set him on a path of earnest research to discover how the DNA code made the building blocks of proteins. This experiment drew on the work of fellow Cornell biochemist Dr. Elizabeth B. Keller, who would continue to have an impact on his discoveries.

Back at Cornell, Holley and his team conducted detailed biochemical experiments over the course of years in order to work out the complete molecular sequence of transfer RNA, which was a key discovery

---

9   U. L. Rajbhandary, R. J. Young, and H. G. Khorana, "Studies on Polynucleotides. XXXII. The Labeling of End Groups in Polynucleotide Chains: The Selective Phosphorylation of Phosphomonoester Groups in Amino Acid Acceptor Ribonucleic Acids," *Journal of Biological Chemistry* 239 (1964): 3875–84.

10 K. Minamoto, M. H. Caruthers, B. Ramamoorthy, J. H. van de Sande, N. Sidorova, and H. G. Khorana, "Total Synthesis of the Structural Gene for the Precursor of a Tyrosine Suppressor Transfer RNA from Escherichia coli. 3. Synthesis of Deoxyribopolynucleotide Segments Corresponding to the Nucleotide Sequence 27-51," *Journal of Biological Chemistry* 251: 587–98.

in explaining the synthesis of proteins from messenger RNA.[11] Holley's findings here were again facilitated by the work of Dr. Elizabeth Keller, who, with a student, proposed a cloverleaf shape—with a stem and three loops—for the transfer RNA molecule. This went far to explain the spatial relationships that facilitated the assembly of proteins from the codes in RNA.[12] Holley mentioned Keller twice in his Nobel lecture, which included a diagram of her cloverleaf tRNA model. Holley further recognized her important contribution by sharing some of his prize winnings with her.[13]

The method devised in Holley's lab was later modified to detect the RNA and DNA structures in various viruses. Holley's work was a precursor to the methods used in the twenty-first century to test for viruses causing COVID, influenza, and other infectious diseases.

In his Nobel lecture, Holley cheerfully credited the sabbatical, a school year free from teaching duties that can be devoted to travel, writing, or—as in Holley's case—a stint as a guest researcher in someone else's lab. "That then is our story. . . . It all followed quite natu rally from taking a sabbatical leave. I strongly recommend sabbatical leaves."[14]

In 1968, Holley moved to the Salk Institute for Biological Studies in La Jolla, California, and two years later he became the founding director of the Salk Cancer Center, one of the first National Cancer Institute–designated basic research cancer centers in the United States. Year after year, the Salk Cancer Center is the recipient of millions of dollars of funding, including specific money for lung cancer research, since it is the most common cause of cancer death in America. During

---

11 R. W. Holley et al., "Structure of a Ribonucleic Acid," *Science* 147 (1965): 1462–65.

12 J. S. Loehr and E. B. Keller, "Dimers of Alanine Transfer RNA with Acceptor Activity," *Proceedings of the National Academy of Sciences of the United States of America* 61, no. 3 (1968): 1115–22.

13 F. Burkhart, "Dr. Elizabeth Keller, 79, Dies; Biochemist Helped RNA Study," *New York Times*, December 28, 1997.

14 R. W. Holley, "Alanine Transfer RNA," December 12, 1968, NobelPrize.org, nobelprize.org/prizes/medicine/1968/holley/lecture/.

Holley's lifetime, the research at Salk and other cancer centers made no impact on the rising death rates due to lung cancer. US lung cancer death rates increased from 45 per 100,000 population in 1970 to 54 per 100,000 in 1993, the year of Robert Holley's death at the age of seventy-one from lung cancer.[15]

———

Nobel organization records show that none of these 1968 laureates were nominated for the Prize in Medicine; instead, all were nominated for the prize in chemistry, there being no prize category of "biochemistry." On the chemistry side that year, the Nobel Prize was given to Lars Onsager for elucidating a universal law of thermodynamics. This was an instance of the special rule of the Nobel Foundation that reads: "Work done in the past may be selected for the award only on the supposition that its significance has until recently not been fully appreciated."[16] Onsager's 1929 publication of his phenomenal discovery was so ahead of its time that it garnered little attention for the next three decades. The recognition was long overdue, and it is probable that the Nobel Committee did not want to dilute his recognition. The work of the three biochemists on related aspects of the genetic code fit properly into a shared medicine prize.

---

15 "Lung Cancer Death Rates, 1950 to 2002," Our World Data, ourworldindata.org/cancer. Data collected by World Health Organization.

16 Claesson, S. Award Ceremony Speech, www.nobelprize.org/prizes/chemistry/1968/ceremony-speech

19

Viruses Everywhere

The 1969 Nobel Prize in Medicine went to Max Delbrück, Alfred Hershey, and Salvador Luria for their discoveries concerning the replication mechanism and genetic structure of viruses.

Max Ludwig Henning Delbrück was born in Berlin in 1906, the youngest of seven children in a well off family in a neighborhood surrounded by extended family. He was interested in science from a young age and seized upon astronomy to study, specifically because it was something no one else was doing—he could engage in it without competition from his siblings and without comparisons to his variously talented cousins and eminent older relatives.

Delbrück recalls the hunger and cold of the WWI years. "An hour after any meal you were hungry again." Heating coal was not available for several winters. "My brother and I became great experts in rope jumping to keep warm; between doing our schoolwork every ten minutes we jumped up."[1] Fully 75 percent of the young men in the family, including his eldest brother, were killed in the war.

Delbrück studied astronomy at university and then shifted to physics, attracted by the new field of quantum mechanics. He did

1 "Max Delbrück (1906–1981) Interviewed By Carolyn Harding July 14–September 11, 1978," Archives of the California Institute of Technology, oralhistories.library.caltech.edu/16/1/OH_Delbrück_M.pdf.

experiments on the deflections of gamma rays, and a phenomenon he discovered was later named "Delbrück scatter."[2] Delbrück became intrigued with the possibility that the gene could be described in terms of quantum mechanics, proposing that it was the basic unit of life. His paper on this concept, coauthored with two others, was instrumental in launching the new field of molecular biology.[3]

In the 1930s, the National Socialist Party rose to power with Hitler, and the Nazis instituted controls at every level of life, including academic. In 1937, when Delbrück sought approval for a university lectureship position, the Nazis required a process of political vetting. Delbrück described the vetting this way, "...you had to go to a *Dozentenakademie*, an indoctrination camp, which was quite a fascinating thing. A 'free' discussion group, you know, where you got lectures on the new politics and the new state. So we had 'free' discussions, and after three weeks of 'free' discussions they decided whether you were sufficiently politically mature to become a lecturer at the University." Delbrück figures he was too incautious in the free discussions, because he didn't pass but was advised he could try again, which he did. "But still I must have shot my mouth off. It must have been transparent that I wasn't in great love with the new regime, so I don't know whether I was officially informed that I wasn't mature enough, or whether they just didn't answer my letters."[4]

---

2 M. Delbrück, "Zusatz bei der Korrektur" [Additions in the corrections], appendix to L. Meitner and H. Rosters, "Lleberstreuung kurzwelliger y-strahlen" [About the scattering of short-wave rays], *Zeitschrift fur Physik* 84, no. 137 (1933): 144. Delbrück scatter is described further in F. Rohrlich and R. Gluckstern, "Forward Scattering of Light by a Coulomb Field," *Physical Review* 86 (1952): 1–9.

3 N. W. Timofejew-Ressowski, K. G. Zimmer, M. Delbrück, *Über die Natur der Genmutation und der Genstruktur* [About the nature of the gene mutation and the gene structure] (Berlin: Weidmann, 1935).

4 "Max Delbrück (1906–1981) Interviewed By Carolyn Harding July 14–September 11, 1978," Archives of the California Institute of Technology, oralhistories.library. caltech.edu/16/1/OH_Delbrück_M.pdf.

Nazi Control of German Universities

Hitler feared the independent thinking that was famously fostered by the nation's more liberal academic institutions, and his attack on the universities started shortly after he was appointed chancellor in 1933. Bernhard Rust, Hitler's cultural affairs minister, decreed that students and teachers must greet each other with the Nazi salute. Reliable Nazis were appointed to be university rectors, empowered to carry out the Nazi aims of ethnic reorientation. The idea was to eliminate anything Jewish or non-German from their ranks, studies, research projects, and publications.[5] This was done through the National Socialist German Lecturers League (NSDDB) an organization under the Nazi Party by order of Deputy Führer Rudolf Hess. It is a gruesome fact that the leadership ranks in the Lecturers League were dominated by the medical faculty. Any lecturers who were Jewish, known liberals, or Social Democrats were dismissed. Many more fled before they were fired, including non-Jewish sympathizers and others who quietly disagreed with Nazi policies.

Frankfurt University was the first institution attacked, because it was the most liberal. An eyewitness account by Peter Drucker, an Austrian who was a lecturer in economics, tells the story: "The new Nazi commissar wasted no time on the amenities. He immediately announced that Jews would be forbidden to enter university premises and would be dismissed without salary on March 15; this was something that no one had thought possible despite the Nazis' loud antisemitism. Then he launched into a tirade of abuse, filth, and four-letter words such as had been heard rarely even in the barracks and never before in academia. . . . [He] pointed his finger at one department chairman after another and said, 'You either do what I tell you or we'll put you into a concentration camp.'"[6]

5   S. P. Remy, *The Heidelberg Myth: The Nazification and Denazification of a German University* (Cambridge, MA: Harvard University Press, 2002), 13, 72.

6   M. Stern Strom, *Facing History and Ourselves: Holocaust and Human Behavior* (Brookline, MA: Facing History and Ourselves National Foundation, 1994).

No one knew how long the Hitler terror would last, and many Germans in academia decided to stick it out despite their disagreement with the regime. Max Delbrück hoped for this, but his poor performance at the indoctrination school meant that he could not even secure an unpaid lecturer position, and so he accepted a Rockefeller Foundation offer to study in Bristol, England. Much of his family stayed behind, and they were devastated by the war. In WWII, eight members of Delbrück's immediate and extended family were active in the resistance, and seven of them were among 4,980 people executed in a roundup following the failed assassination attempt on Hitler on July 20, 1944. The husbands of two of Max's sisters were killed by marauding soldiers in the chaos of the last days of the war. Max's brother Justus was in a Nazi prison awaiting trial when the war ended and, after a brief liberty, was picked up by Russian occupiers who wanted him to testify at the Nuremberg trails regarding Nazi atrocities. Justus died of diphtheria while he was held in a Soviet camp.

From Bristol, Delbrück went to America. At the California Institute of Technology (Caltech) in Pasadena, California, he started research on viruses that infect bacteria (bacteriophages, or "phages" for short), considering this a much easier way to study genetics than breeding flies. Here he discovered that virus infection takes place in one step, rather than exponentially, as is seen in higher organisms.[7] The virus enters the bacterium, and the host's cellular machinery is used to assemble viral replicates, which then burst the bacterium so the virus particles are released to infect other cells. His experiments were carried out on culture plates with lawns of bacteria growing on them. When the phages were introduced, each was found to affect a small area of bacteria, leaving a blank hole in the lawn about 1 mm in diameter. By holding the plate up to the light and counting the tiny holes, one could count how many phage particles had infected the bacteria.

---

7   E. L. Ellis and M. Delbrück, "The Growth Of Bacteriophage," *Journal of General Physiology* 22, no. 3 (1939): 365–84.

Delbrück was so excited by these results that he ran around the labs at Caltech showing off his culture plate. He was greeted with many murmurs of appreciation and congratulations, until he showed it to the wife of the department chairman. She admitted she saw nothing, and let Delbrück know that no one else had seen anything either. He responded that what had happened was just like "The Emperor's New Clothes" and was surprised the Americans had never heard of the Hans Christian Andersen story. When he dashed off to a bookstore to get a copy, he was further surprised that the clerk didn't know the difference between stories by Hans Christian Andersen (stories that Andersen made up) and the Grimm Brothers (a collection of traditional folk tales).

War broke out in Europe two years after Delbrück had come to the US, so his return to Germany was neither desired nor possible. He met and married Mary Bruce at Caltech before the Rockefeller Foundation subsidized a position for him at Vanderbilt University in Tennessee.

At Vanderbilt, Delbrück was visited by colleagues in the new field of phage research, including future Nobel cowinners Alfred Hershey and Salvador Luria. With a few others, they formed the "phage group," meeting every summer at the labs and symposia centers of Cold Spring Harbor in Massachusetts. On his first visit, Delbrück was excited to see that Luria had already been working with an electron microscopist to visualize a virus. Due to wartime restrictions on the free exchange of information with Germany, the group was unaware that H. Ruska had already published the first ever electron microscope picture of a virus.[8]

---

8   H. Ruska, "Die Sichtbarmachung der bakteriophagen Lyse im Uebermikroskop" [The visualization of bacteriophage lysis under the microscope], *Naturwissenschaften* 28 (1940): 45–46; H. Ruska, "Ueber ein rreues bei der bakteriophager: Lyse auftretendes Formelement." [About a true in bacteriophage: Lysis occurring form element], *Naturwissenschaften* 29 (1941):367–68.

The Principle of Limited Sloppiness

At a meeting on the topic of photoreactivation, a phenomenon whereby bacteria and phage come back to life after inactivation by ultraviolet rays, Delbrück made a casual remark that became well known. Here's how he remembers it:

"Many scientists had irradiated bacteria and phage with ultraviolet light, including Luria, myself, Dulbecco, and so on and so forth, and had measured survival rates. It turns out that if you measure the survival in the presence of daylight, then you get entirely different values than when you measure survival in the dark or in red light. The reason that it hadn't been discovered was because whoever had done the measurements had done them very carefully under controlled conditions, always the same light. Both Kelner and Dulbecco had done the experiments in a little more sloppy way, sometimes putting the plates here, and sometimes putting the plates there, sometimes having the water bath near the window, and sometimes not near the window, and then noting that there was something grossly different. So in introducing this little symposium, I said it shows the usefulness of limited sloppiness. If you are too sloppy, then you never get reproducible results, and then you never can draw any conclusions; but if you are just a little sloppy, then when you see something startling, you say, 'Oh, my God, what did I do, what did I do different this time?' And if you really accidentally varied only one parameter, you nail it down, and that's exactly what happened in both of these cases. So I called it the 'Principle of Limited Sloppiness.'"[9]

These variances led to the discovery that the section of DNA damaged by ultraviolet is repaired by a light-activated enzyme, explaining why exposure to varying amounts of light would cause differing results.

---

9   Interview with Max Delbruck (1978), p. 76-77. Oral History Project, California Institute of Technology Archives, Pasadena, California.

Max Delbrück's personality and intentions are credited with creating a congenial atmosphere in the phage group and the free exchange of information, much in contrast to the competitions and secrecies in later years of genetic code research. Delbrück described the ambience at Cold Spring as intimate and creative in an unpressured way. To facilitate comparing results and building on each other's work, the group agreed on the exact viruses they would all study (T1-4) on just one type of bacteria (*E. coli*).

In 1942, Delbrück collaborated with Salvador Luria to publish a study demonstrating that most mutations in bacterial DNA were spontaneous and not caused by pressures or exposures in the environment.[10] The Luria-Delbrück experiment showed that mutations that made the *E. coli* bacteria resistant to the harmful effects of T1 bacteriophage (virus) existed in the bacteria population prior to exposure to T1 and were not induced by adding T1. In other words, mutations are random events that occur whether or not they prove to be useful. After exposure to the virus, most bacteria sensitive to T1 died, but the unharmed T1-resistant bacteria propagated, and their progeny, which had inherited T1 resistance, dominated future bacterial generations. This demonstrated that a spontaneous mutation could combine with Darwinian survival of the fittest, even though the mutation did not come about because of outside influences. This finding influenced an entire generation of DNA researchers to conclude that all life originated from spontaneous, chance mutations and developed stepwise into more and more adaptive organisms because successive lucky spontaneous mutations happened to give survival advantages.

It has since been discovered that outside influences that do not confer survival advantage can cause changes to DNA in which the defective DNA survives and is transmitted to future generations. An example is diethylstilbestrol (DES), a synthetic form of the female hormone estrogen that was given to pregnant women between 1940 and

---

10 S. E. Luria and M. Delbrück, "Mutations of Bacteria from Virus Sensitivity to Virus Resistance," *Genetics* 28, no. 6 (1943): 491–511.

1971 to prevent miscarriage. Not only did DES not help the pregnancy, it caused increased cancer risk in women who took it, as well as in their sons and daughters. Ongoing research shows that third-generation offspring have higher rates of infertility, although higher cancer risk in the third generation has not yet been established.

Delbrück directed a course in phage research for ten summers at Cold Spring. He left the field when he felt that younger scientists would capably carry it forward, turning to the study of the fungus *Phycomyces*. He investigated what caused its attraction to light, aversion to barriers, and antigravity growth patterns. He stayed at Caltech for the remainder of his career, and upon retirement he was made professor emeritus so he could continue his research.

Delbrück strongly disagreed with the way the development of science is represented in most textbooks, as an orderly progression from hypothesis via experiment to conclusion.

> The progress of science is tremendously disorderly, and the motivations that lead to this progress are tremendously varied, and the reasons why scientists go into science, the personal motivations, are tremendously varied. I have said what I have to say about that in the Beckett lecture, at least one particular point that seems to be missed; that science is a haven for freaks, that people go into science because they are misfits, and that it is a sheltered place where they can spin their own yarn and have recognition, be tolerated and happy, and have approval for it.[11]

Max Delbrück died at age seventy-four in 1981 after a short struggle with multiple myeloma, a bone marrow cancer.

———————

11 M. Delbrück, "Homo Scientificus According to Beckett," in *Science, Scientists, and Society*, W. Beranek Jr., ed. (Tarrytown-on-Hudson, NY: Bogden & Quigley, 1972), 132–52.

Salvador Edward Luria was born Salvatore Luria in Turin, Italy, in 1912. In his youth, he saw the rise of Fascism and was exposed to teachers who resisted the trend. He graduated from medical school in 1935, but after eighteen months of compulsory army service as a medic, he knew he did not have a passion for clinical medicine. He pursued radiology but was disappointed to find that the clinical radiologists knew almost nothing of the physics that made radiation possible, and so he turned to a more classical study of physics. This led him to the emerging field called biophysics.

About that same time, Luria learned about bacteriophages (phages)—a large group of viruses that prey on bacteria. He describes his first experience with phages: "One drop of a one-to-a-billion dilution of a phage solution could completely dissolve a culture of bacteria in a few hours! One phage virion attacks one bacterium, produces a hundred virions, these attack a hundred bacteria, and so on until after a few rounds of attack there are no bacteria left."[12] When Luria became aware of Max Delbrück's theory of DNA being the basic unit of life, he thought that bacteriophages would be an excellent research organism with which to explore the theory.

Anti-Semitism had no place in Italian society before Italy's alignment with Hitler,[13] but by the mid-1930s life in Italy became untenable for anyone with Jewish heritage. As a Jew, Luria found no opportunities in Italy, so he moved to Paris, while the family he left behind was well hidden by sympathetic neighbors. (They all survived the Nazi occupation of Italy.) In Paris, Luria studied in a fellowship program at the Institute of Radium under the supervision of physicist Fernand Holweck.

---

12 S. E. Luria, *A Slot Machine, a Broken Test Tube: An Autobiography*, (New York: Harper and Row, 1984).

13 Gene Bernardini. "The Origins and Development of Racial Anti-Semitism in Fascist Italy, *Journal of Modern History 49, no. 3 (1977): 431–53.*

As the German army approached Paris in June 1940, Luria resolved to flee. The night before he left, Eugène Wollman, phage researcher from the Pasteur Institute, took Luria to his home and made him listen to a Beethoven recording "to restore my faith in humanity," recalled Luria in his 1984 autobiography.[14] Luria rode a bicycle out of Paris, hiding in the homes and barns of sympathizers along the way. As he cycled through the French countryside on his way to Marseilles, he twice came under strafing from German airplanes, both times with no harm done. In Marseilles he sought out the American embassy, where he waited with the desperate horde seeking immigration visas. He passed the inquisition-style interview into his personal, political, and sexual conformities, and he was at last granted a visa. Others were not so lucky; Luria described seeing suicides on the steps of the embassy. He took a train to Lisbon, where he had a two-week wait for berthing on a Greek ship headed for New York. He arrived in the US with the suit on his back and fifty-two dollars in his pocket. His colleagues in Paris were slain by the Nazis: in December 1941, the physicist Holweck was arrested, tortured, and murdered for assisting downed British parachutists; in December 1943, Wollman and his wife Elisabeth were arrested, deported, and murdered for being Jewish.

In New York, Luria was granted a Rockefeller Foundation fellowship at Columbia's College of Physicians and Surgeons thanks to a letter of recommendation from his countryman Enrico Fermi. When Fermi won the 1938 Nobel Prize in physics, he took the opportunity to travel directly from Stockholm to America to escape racial oppression in Italy, so he completely understood Luria's situation.

Luria met Max Delbrück in December 1940, and the following summer, the two did their first investigations into how phage multiply within bacteria. They found that when a phage strain infects a bacterium, a different phage strain is unable to attack the same bacterium. In 1943, Luria joined the faculty at Indiana University, where he

---

14 S. E. Luria, *A Slot Machine, a Broken Test Tube: An Autobiography*, (New York: Harper and Row, 1984).

demonstrated that bacteria mutated spontaneously into phage-resistant forms. He sought collaboration with Delbrück in order to use these experiments to calculate bacterial mutation rates. Luria's first graduate student was the young Jim Watson, whom he set to work on experiments with photoreactivation, where phage killed by ultraviolet light reconstituted themselves. While in Indiana, Luria met and married Zella Hurwitz, and he became a US citizen, taking the opportunity to change his first name to Salvador and to add a middle name, Edward.

The beginning of Luria's political activism consisted of campaigning for the left-wing Progressive Party in the 1948 elections. The Progressives backed Henry Wallace for US president on a platform of desegregation, along with the expansion of welfare, establishment of a national health insurance system, nationalization of the energy industry, and conciliation with the Soviet Union. Luria became active in the University Teachers Union and the American Labor Party. These activities were not well received in the conservative environment of Indiana University, which Luria suspected as the reason Indiana failed to counter an offer from the University of Illinois at Urbana.

Luria moved to Illinois, where he continued his work and discovered that bacteria are not always entirely helpless in the face of phage infection: He learned that bacteria make enzymes that modify the expression of phage DNA, restricting viral replication. While at Illinois, he authored an authoritative text on virology that went through three editions; however, he was like Delbrück in generally despising textbooks. He said, "I have a nasty suspicion that a good deal of traditional subject matter is kept in textbooks because it provides convenient if meaningless quiz questions."[15]

In Illinois, Luria continued his political activism, joining the teachers union, publicly opposing legislation that promoted McCarthyism, helping to author a resolution on biological warfare that was later

---

15 S. E. Luria, *A Slot Machine, a Broken Test Tube: An Autobiography*, (New York: Harper and Row, 1984).

adapted by the United Nations, and signing on to Linus Pauling's famous petition against nuclear bomb testing.[16]

In 1959, Luria moved to MIT in Boston and turned his research efforts to the cell membranes of bacteria and the substances that bacteria make to protect against competing bacteria. Eventually he would head a cancer research center at MIT, out of which came some later Nobel laureates. Boston was more conducive to Luria's political activism, and the first thing he did was to organize an ad in the *New York Times* signed by sixty MIT and Harvard faculty to oppose the Bay of Pigs invasion of Cuba by the US. This was followed by another ad opposing President Kennedy's new civil defense program, which many scientists considered nothing more than an effort to whip up anti-Soviet hysteria. Luria helped to form the Boston Area Faculty Group on Public Issues, an organization through which he continued activism for many years. He gave antiwar speeches and marched in protests at the onset of US involvement in Vietnam in 1963, and he supported antidraft activists. He opposed the building of new atomic power plants in the 1970s, and he spoke against the 1982 Israeli invasion of Lebanon. Luria described himself as a socialist committed to "political radicalism."

Salvador Luria died of a heart attack in 1991 at the age of seventy-eight.

The third phage scientist to share the 1969 prize was Alfred Day Hershey, who was born in 1908 in Owosso, Michigan. He attended Michigan State College for his bachelor's of science and PhD degrees and then joined the Washington University School of Medicine, teaching and researching in bacteriology. His initial discoveries concerned the reproductive characteristics of bacteria. Like so many others, Hershey soon became interested in genetic material. In the early 1940s

16 G. Hill, "2,000 Join Pauling in Bomb Test Plea; He Lists Backers of World Ban— Other Scientists Dispute Radiation Views," *New York Times*, June 4, 1957.

his attention turned to bacteriophages, as these viruses were the simplest form of genetic material that had yet been found. It was known that when viruses infect bacteria, they inject their genetic material then use the machinery of the bacterial cell to assemble more virus particles; but what exactly was this viral "genetic material"?

Previous studies had demonstrated that genetic information was not in the fats or sugars of a cell. There was speculation the genes were protein, and some thought it was RNA. A simple experiment was reported in 1944 that was trying to answer a different but related question: How did bacteria swap DNA among themselves? In this experiment, bacterial extracts had all of their fats, sugars, proteins, and RNA removed, but the extract could still transmit genetic information. It was only when researchers destroyed the DNA that genetic information could not be preserved. These experiments supplied resounding evidence that DNA was the carrier of genetic information, but the researchers (Avery et al.) did not push their findings at scientific meetings, and so the discovery went largely unnoticed. With more self-promotion it might have won a Nobel Prize.[17]

Max Delbrück invited Al Hershey to Nashville in 1943 to discuss using phage to pin down the genetic material. Delbrück made these notes on his guest: "Drinks whiskey but not tea. Simple and to the point . . . Likes to be independent."[18] Along with Salvador Luria, they became the leadership of the phage group that met yearly at the Carnegie-funded Cold Spring Harbor labs. By 1950, Hershey moved east and became a full-time staff member in genetics at Cold Spring.

---

17 O. T. Avery, M. C. MacLeod, M. McCarty, "Studies on the Chemical Nature of the Substance Inducing Transformation of Pneumococcal Types: Induction of Transformation by a Deoxyribonucleic Acid Fraction Isolated from Pneumococcus Type III," *Journal of Experimental Medicine* 79, no. 2: 137–58.

18 F. W. Stahl, *Alfred Day Hershey 1908–1997: A Biographical Memoir* (Washington, DC: National Academy Press, 2001).

In 1952, Hershey and Martha Chase devised a simple experiment.[19] First, they labeled phage DNA by spiking it with radioactive phosphorus, since phosphorus is a molecule that only the DNA has. Then, they labeled phage protein by spiking it with radioactive sulfur, a molecule unique to protein. These radioactive labels allowed them to trace the locations of phage DNA and proteins throughout the experiment. Next, they mixed the labeled phage with some bacteria and allowed some time for the phage to attach to the bacteria and infect them. During this process, they knew, the phage transferred its genetic material into the bacteria. They put the mixture in a blender in order to shake the phage loose from the bacteria. Then, they put the mix in a centrifuge; the phage particles settled at the bottom layer, and the bacteria rose to the top layer.

They found all of the radiolabeled phosphorus—meaning the phage DNA—in the top (bacterial) layer. All of the radiolabeled sulfur—the phage protein, was in the bottom (phage) layer. They concluded that phage transferred its DNA into the bacteria, while phage protein remained with the phage, outside the bacteria. This work provided the strongest support yet for the hypothesis that DNA is the conveyor of genetic information, and it was the primary discovery that earned Hershey the Nobel Prize.

Who Was Martha Chase?

Martha Chase was Al Hershey's coresearcher on the landmark Hershey-Chase experiment. She was born in 1927 in Cleveland, Ohio, and received a bachelor's degree from the College of Wooster in 1950, then worked as a research assistant at Cold Spring. Chase's name appeared as second author on their famous paper of 1952, for which Hershey would win the Nobel. The next year, Chase moved to Oak Ridge National Laboratory in Tennessee and later

---

19 A. Hershey and M. Chase, "Independent Functions of Viral Protein and Nucleic Acid in Growth of Bacteriophage," *Journal of General Physiology* 36: 39–56.

she went to the University of Rochester, but throughout the 1950s, she continued to participate in the summer phage meetings at Cold Spring. She then attended the University of Southern California for her PhD, which she earned in 1964. Hershey did not cite their landmark paper in his 1969 Nobel lecture, nor did he even mention her. Due to personal issues, she dropped out of her scientific career. Chase suffered from premature dementia, and she died in 2003.[20]

When asked what his idea of happiness would be, Hershey replied, "To have an experiment that works, and do it over and over again."[21] By the time of the 1969 Nobel Prize announcement, Alfred Hershey was the only one of the three laureates who was still active in phage research.

Albert Hershey died of cardiopulmonary failure at the age of eighty-eight in 1997.

20 B. Bibel, "Martha Chase," The Bumbling Biochemist, March 14, 2018, thebumblingbiochemist.com/wisewednesday/martha-chase/.

21 F. W. Stahl, ed., *We Can Sleep Later: Alfred D. Hershey and the Origins of Molecular Biology* (Cold Spring Harbor, NY: Cold Spring Harbor Laboratory Press, 2000).

20

Brain Chemicals

The 1970 Prize was shared by Ulf von Euler, Bernard Katz, and Julius Axelrod for their independent discoveries about biochemicals of the nervous system.

For most Nobel laureates, the awards ceremony in December is a novel experience. Not so for Ulf von Euler, who attended the festivities as a true insider. Born in 1905 in Stockholm, von Euler's father was Hans von Euler-Chelpin, previously a Nobel winner in chemistry; no doubt Ulf had attended his father's awards ceremony. Ulf's godfather was Svante Arrhenius, another previous chemistry laureate. In the 1920s, Ulf von Euler attended medical school at the Karolinska Institute, the home of the committee for the Nobel Prize in Medicine. Von Euler studied abroad in the labs of medicine Nobel winners Archibald Vivian ("A.V.") Hill, Henry Hallet Dale, Corneille Heymans, and Bernardo Houssay. From 1953 to 1960, von Euler was a member of the Nobel Committee for Physiology or Medicine, and from 1961 to 1965, he served as secretary of the committee. In these capacities, he recommended the Nobel Prize recipients. In 1965, he was appointed chairman of the board of the Nobel Foundation.

Von Euler's biochemical research can be summed up as "exciting"—he studied the various substances in the body that stimulate, arouse, and agitate. His first significant discovery was in 1931, when he identified substance P, which is released by tissues whenever there is a stressor, such as an injury, illness, or other threat to bodily survival.[1] Substance P amplifies the body's reactions for defense and repair, and it transmits pain sensation through the central nervous system. Receptors to substance P are specialized surfaces on the cells on which it acts, and they are found throughout the body, including on nerves, capillaries, lymphatics, stem cells, and white blood cells. Despite the knowledge that levels of substance P are elevated in many illnesses, measurement of it has not been useful to diagnose specific diseases. Its roles in pain and inflammation are known, but attempts to design medications related to substance P, such as substance P inhibitors to treat pain, have failed.

In 1936 von Euler detected something secreted by the genital glands of males that causes blood vessels to dilate and muscle fibers to contract. This action opens the penile vessels so they can fill with blood, inducing engorgement to achieve an erection. Von Euler named this substance prostaglandin, since he had isolated it from the prostate gland.[2] Subsequent research found that prostaglandins are a family of related chemicals and are produced in nearly all tissues of the body, but the name has stuck. Today prostaglandins are used for inducing labor, treating certain heart defects in newborns, stomach ulcers, irritable bowel syndrome, and glaucoma. In an echo of its original discovery, prostaglandin is used to treat erectile dysfunction, in which case it is self-injected into the penis.

The discovery that earned von Euler recognition by the Nobel Committee was his 1946 identification of noradrenaline, also called norepinephrine.[3] It is the key chemical that transmits nerve impulses

1   U. S. von Euler and J. H. Gaddum, "An Unidentified Depressor Substance in Certain Tissue Extracts," *Journal of Physiology* 72, no. 1 (June 1931): 74–87.

2   U. S. von Euler, "On the Specific Vaso-Dilating and Plain Muscle Stimulating Substances from Accessory Genital Glands in Man and Certain Animals (Prostaglandin and Vesiglandin)," *Journal of Physiology* 88, no. 2 (1936): 213–34.

3   U. S. von Euler, "The Presence of a Sympathomimetic Substance in Extracts of Mammalian Heart," *Journal of Physiology* 105 (1946): 38–44.

to trigger a physical response, which prepares the body to either fight or flee a stressful scene. When a person experiences a dangerous situation, their noradrenaline level surges, causing muscles to contract more forcefully. This includes the heart muscle, which constricts blood vessels to raise blood pressure. Noradrenaline is released by the nerve endings even when a person only contemplates a life-threatening situation in calm surroundings. This is illustrated on a graph showing von Euler's measurements of noradrenaline levels in parachutists on jump-training days compared to routine days. Today noradrenaline is used clinically as a means of maintaining blood pressure in certain types of shock, for example, circulatory collapse from overwhelming infection.

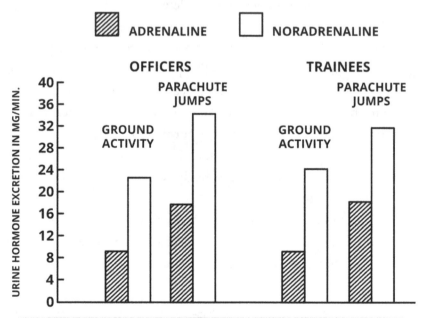

Modified from U. von Euler, Adrenergic Neurotransmitter Functions, Nobel Lecture, December 12, 1970.

Sweden in WWII

Throughout history the Swedish state has had to find ways to be relatively assertive and at other times conciliatory in order to maintain independence from its two powerful neighbors, Russia and Germany. Sweden's complicated compromise in WWII was to declare itself neutral in 1939, while it aided both sides in different ways. After Nazi occupation of Denmark and Norway in 1940, Sweden allowed the Nazis to use its rail lines to secure iron ore crucial to the manufacture of tanks, armaments, and airplanes, knowing the Germans also packed the rail cars with armed soldiers on their way to Norway, Finland, and Russia. Official Swedish print media was tightly controlled so as not to appear to favor one side or the other, but the Nazis broadcasted radio from the Prussian city of Königsberg (now Kaliningrad in present-day Russia), and the Allies broadcast via the BBC from London. Ulf von Euler's second wife was the Swedish countess Dagmar Carola Adelaide Cronstedt, who worked at Radio Königsberg reading Nazi propaganda to neutral Sweden. Sweden is a northern Germanic country, and Swedes have many blood connections with Germany. Prominent Swedes with Nazi ties during the war included Ingvar Kamprad, the founder of IKEA, and Walther Sommerlath, an arms manufacturer and the father of the current Swedish queen, Silvia.

There were also considerable efforts to aid the Allies. For example, in 1940, Swedish mathematician Arne Beurling broke the code of the Nazi ciphering machine, allowing the Swedes to supply the Allies with crucial intelligence. Sweden provided refuge for thousands of Jews from occupied Scandinavian countries during the war, and many Swedish individuals were active in collecting intelligence. In the last two years of the war, Swedish airbases were used by the Allies.

Unfortunately von Euler did not get to share his joy of the prize with his father, who had died six years earlier.

Ulf von Euler died from complications of surgery in 1983 at the age of seventy-eight.

————

Bernard Katz was another of the cowinners of the 1970 prize. He was born in 1911 in Leipzig, Germany, where his parents had immigrated to escape anti-Semitism in Russia. By 1929, as Katz graduated gymnasium (high school), the National Socialist German Worker's Party was a rising influence in German society, and Hitler soon became its undisputed leader. Like some other especially alert German Jews, Katz considered emigrating, but he became convinced it was better to get through medical school first to make himself a more attractive visa applicant whenever he did decide to leave the country.

Hitler was appointed chancellor of Germany in January 1933, when Katz was in his third year in medical school. The following month, a suspicious fire at the German parliament building became the excuse for the emergency suspension of constitutional protections and the beginnings of a police state. In March, the Nazis opened the first concentration camp outside the city of Dachau. By midyear, there had been political purges, book burnings, and violence in the universities. That same year, Katz was awarded the Siegfried Garten Prize for excellence in physiology for his research into the mechanisms of nerve transmission. However, the prize was not presented in the usual public way, for by then, the Nazi program of scrubbing all traces of Jewish contributions to science was well underway. Katz's department head had to give him the prize privately. Katz completed his state exams and graduated medical school in November of 1934.

Like von Euler, Katz was invited to research in the lab of the world's premier biophysicist and Nobel laureate, A. V. Hill, in England. For Katz this was not only a treasured scientific opportunity: the invitation was extended to place him beyond the reach of the Nazis. Hill was unlike most of his scientific colleagues, who were choosing to stay silent on the growing terrors of Nazism. As early as 1933, Hill appealed to the world's scientific community to publicly endorse a moral stance against the events unfolding in Germany. Hill became president of the Academic Assistance Council (AAC), which helped hundreds of scientists to flee the Nazis, sixteen of whom would become Nobel Prize winners.[4]

Excerpts from A. V. Hill's "International Status and Obligations of Science"

"The facts are not in dispute. Apart from thousands of professional men, lawyers, doctors, teachers, who have been prevented from following their profession, apart from tens of thousands of tradesmen and workers whose means of livelihood have been removed, apart from 100,000 in concentration camps, often for no cause beyond independence of thought or speech, something over 1,000 scholars and scientific workers have been dismissed, among them some of the most eminent in Germany."

"We cannot take the freedom, so slowly and hardly won, as a birthright: we must see to it that neither race, nor opinion, nor religious belief, nor the advocacy of theories unpopular perhaps at the moment shall cause disinterested able men to be deprived of the means of carrying on their work."

". . . I can imagine that Homo sapiens may ultimately destroy, by his irreconcilable folly, all he has built up."

"Mankind's amazing intellectual achievement in understanding and controlling the forces of Nature may be neutralized by the domination of his intellect by his passions, by his emotional inability to realise, what must be obvious to his intellect alone, the demands of a common humanity."[5]

---

4 *Boneheads & Brainiacs.*
5 A. V. Hill, "International Status and Obligations of Science," *Nature* 132, no. 3347 (1933): 952–54.

Nazi suppression even reached Katz in England when he submitted his manuscript on the mechanisms of nerve transmission to the German journal *Ergebnisse der Physiologie* in 1937. It was rejected for the lack of an Aryan coauthor. The paper was ultimately published in the *Oxford Review*, and it satisfied Katz's PhD requirements.

While in London, Katz also worked with John Carew Eccles, who would become a Nobel winner in 1963. When Eccles returned to his native Australia, he invited Katz to work with him there. In 1939, Katz was able to escort his parents out of Germany to Australia, where he continued work on the biochemistry of nerve transmission, but Katz's career was put on hold while he served in the Australian army. After the war, Katz returned to England to University College in London, ultimately taking over the department in 1952 when Hill retired.

Katz's research on the mechanics of nerve transmission was preceded by the controversy between the sparkers and the soupers. Charles Sherrington and John Carew Eccles were sparkers, who favored the theory that nerve impulses were transmitted electrically. Henry Dale and Otto Loewi, soupers, thought that a chemical neurotransmitter passed the signal from nerve to nerve. Katz is reported to have been impressed by the intensity of the arguments between the two camps at a public scientific meeting, but also by the friendly personal relations between the protagonists when the meeting was over.[6]

In fact, both methods of nerve transmission occur, purely electrical or by chemical neurotransmitters. Katz's contribution was to clarify the events leading up to the release of neurotransmitters at the nerve endings.[7] His culminating work on the subject, published the year of the prize, brought together the soupers and the sparkers by describing how the release of the chemical neurotransmitter is brought about by

---

6   B. Sakmann, "Bernard Katz: 26 March 1911–20 April 2003," *Biographical Memoirs of Fellows of the Royal Society* 53: 185–202, doi.org/10.1098/rsbm.2007.0013.

7   B. Katz, *The Release of Neural Transmitter Substances* (Liverpool: Liverpool University Press, 1969).

an influx of calcium through channels; these calcium channels, in turn, are opened by the electrical nerve impulse.[8]

Bernard Katz died in 2003 at the age of ninety-two.

———————

The third cowinner in 1970, Julius Axelrod, had a long road to the Nobel Prize. He was born in 1912 in Manhattan, the son of Polish immigrants. He attended New York public schools followed by a year of college at the prestigious New York University (NYU). In a lab at NYU, he suffered eye damage when he attempted to knock a glass stopper out of a bottle of ammonia. The contents had been under pressure and the bottle broke, exploding in his face. Axelrod lost sight in his left eye and from then on wore a distinctive black patch or glasses with one darkened lens.

Axelrod soon ran out of tuition money and had to continue his education at The City College of New York, where he earned a biology degree. He applied for medical school but was not accepted, which he attributed to his less-than-straight-A grades excluding him from the limited quota reserved for Jewish applicants.[9] Axelrod attended night school for his master's degree while working in a government lab, where his job was to validate the vitamin content of commercial foods. He next moved to the lab at a hospital affiliated with NYU, where he worked to discover the biochemical basis of side effects caused by some common painkillers. He found that one drug was broken down in the body to a toxic metabolite and a relatively nontoxic metabolite; the

———————

8 B. Katz and R. Miledi, "Further Study of the Role of Calcium in Synaptic Transmission," *Journal of Physiology* 207 (1970): 789–801.

9 "Dr. Julius Axelrod Oral History 1996 B," interview by M. Flavin, January 26, 1996, National Institutes of Health, Office of History and Stetten Museum, history.nih.gov/display/history/Axelrod%2C+Julius+1996+B.

nontoxic metabolite was what we now know as acetaminophen (brand name Tylenol).[10]

Axelrod realized that, despite his discoveries, he would not get promoted at a university-affiliated lab in the absence of an advanced degree. He thought his chances of career advancement would be better by moving to the National Institutes of Health (NIH), where he initially worked on the metabolism of caffeine. This led him to study other stimulants, such as amphetamines, ephedrine (as in Sudafed and other decongestants), and the hallucinogenic substance mescaline, which is derived from the peyote cactus. He discovered the existence of various liver enzymes that broke down the drugs, which established the groundwork for scientists all over the world to eventually discover that there are many different drug metabolism pathways facilitated by liver enzymes that eliminate drugs and neutralize toxins. Today we understand there are genetic variants of these detoxifying enzymes in the liver, with some people being slow metabolizers and others being fast metabolizers. This can now be tested for in advance of receiving a drug in order to anticipate who will be more likely to suffer side effects, and it has matured into a specialty called pharmacogenomics.[11] Unfortunately, prescribing physicians rarely take advantage of these testing opportunities. In 2018, the FDA made such testing directly available to concerned health care consumers.[12]

Axelrod turned his attention to sedatives, such as morphine and barbiturates, discovering that when the body is continuously exposed,

---

10 B. B. Brodie and J. Axelrod, "The Estimation of Acetanilide and Its Metabolic Products, Aniline, N-acetyl p-aminophenol and p-aminophenol (Free and Total Conjugated) in Biological Fluids and Tissues," *Journal of Pharmacology and Experimental* 94 (1948): 22–28.

11 "Pharmacogenics," National Institutes of Health, nigms.nih.gov/education/fact-sheets/Pages/pharmacogenomics.aspx.

12 US Food and Drug Administration, "FDA Authorizes First Direct-to-Consumer Test for Detecting Genetic Variants That May Be Associated with Medication Metabolism," press release, October 31, 2018, fda.gov/news-events/press-announcements/fda-authorizes-first-direct-consumer-test-detecting-genetic-variants-may-be-associated-medication.

the drugs inactivate the metabolizing enzymes. Furthermore, the continuous exposure of the cells to drugs inactivates the cell membranes' receptors, making them less and less effective to cause sedation or a high. These discoveries were breakthroughs in explaining how an addict needs more and more of the drug.

Despite his important contributions to understanding drug metabolism, at this point Axelrod realized that his lack of a PhD was still preventing his promotion, even at the NIH. At the age of forty-one, he took off a year to attend George Washington University, submitting a great deal of his previous research to satisfy the PhD requirements. Back at the NIH, he formed a new lab at the NIH's National Institute of Mental Health (NIMH) and furthered his investigations of how drugs work, discovering the various aspects of the diverse biochemistry of epinephrine, norepinephrine, serotonin, and dopamine at nerve endings. These neurotransmitters facilitate the transmission of nerve impulse across the gap between nerve endings.

In order to further research serotonin, Axelrod took up the study of lysergic acid diethylamide, also known as LSD, which had been a prescription drug available since 1937. Axelrod said, "...at that time, serotonin was believed to be involved in psychoses because of its structural resemblance to LSD." According to Axelrod, "LSD was then used as an experimental drug by psychiatrists to study abnormal behavior."[13] In the prescribing information for the LSD brand-name product Delysid, the manufacturing company Sandoz wrote that it had three main uses for psychiatrists: first, to produce hallucinatory psychosis in mental patients; second, to experimentally produce psychosis in normal people so as to study mental disorders; and third, for the psychiatrist himself to take so he could experience the world of a mental patient.

---

13 J. Axelrod, "An Unexpected Life in Research," *Annual Review of Pharmacology and Toxicology* 28 (1988): 1–23.

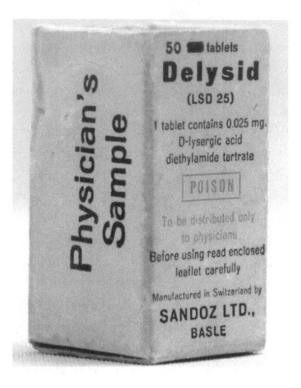

Axelrod experimented on LSD in monkeys, cats, mice and at least one human in order to collect data on the metabolism of the drug.[14] Others at the NIMH were doing human experimentation to document the mental effects of LSD, although many studies were done in the field with prisoners, military personnel, and patients at government-subsidized community clinics. Very little documentation exists on the methods and results, much less on informed consent.[15] It is unclear what the therapeutic effect of LSD was supposed to be. There is no standard medical report of long-term outcomes for patients who were intentionally made to be psychotic by physician-administered

---

14 J. Axelrod et al., "The Distribution and Metabolism of Lysergic Acid Diethylamide," *Annals of the New York Academy of Sciences* 66, no. 3 (1957): 435–44.

15 N. D. Campbell and L. Stark, "Making up 'Vulnerable' People: Human Subjects and the Subjective Experience of Medical Experiment," *Social History of Medicine* 28 no. 4 (2015): 825–48.

LSD, although many horrifying case reports can be found.[16] Eventually, suicides and permanent insanity in experimental subjects and recreational LSD users led to its demise as a medication, and it was pulled off the prescription drug market in 1966.

However, building on the basic research supplied by Axelrod and his colleagues at the government-funded NIMH, drug-development departments of commercial pharmaceutical labs pursued LSD's structural similarity to serotonin with enthusiasm. The first generation of drugs created to specifically affect brain chemicals included Desyrel and Wellbutrin. These were followed by a second generation that included Prozac, Paxil, and Zoloft. The marketing strategy for the latter drugs branded them as selective serotonin reuptake inhibitors, or SSRIs, conveying the idea to physicians and patients that these drugs affected only serotonin out of all the many different neurotransmitters ("selective"), and that their single action was to keep serotonin from being taken up by the cell once it had been used ("reuptake inhibitor"). It turned out that some serotonin-related brain activities are up-regulated by the drugs and some are down-regulated; some brain activities are put into hyperspeed and others are suppressed. What this has to do with reuptake inhibition is uncertain. Furthermore, drugs in this class directly or indirectly affect several other neurotransmitters besides serotonin, including norepinephrine, dopamine, acetylcholine, and GABA; thus, they are definitely not selective.

There are many different subtypes of serotonin receptors, and the drugs affect them each differently. It is now clear that early researchers did not know exactly what they were measuring when they found

16 For more details on individuals affected by experiments with LSD, see, J.S. Ketchum, *Chemical Warfare Secrets Almost Forgotten: A Personal Story of Medical Testing of Army Volunteers* (self-published, 2006); A. McCoy, *A Question of Torture: CIA Interrogation, from the Cold War to the War on Terror* (New York: Henry Holt, 2006); The Fifth Estate, "MK Ultra: CIA Mind Control Program in Canada (1980)," CBC, www.cbc.ca/player/play/1117719107906; C. H. Farnsworth, "Canada Will Pay 50's Test Victims," *New York Times*, November 19, 1992; T. O'Neill and D. Piepenbring, *Chaos: Charles Manson, the CIA and the Secret History of the Sixties* (Portsmouth, NH: William Heinemann Publishing, 2019).

serotonin in various plants and animals and in human brain tissue. In 1979, it was discovered that there were at least two distinct subtypes of serotonin. In the 1990s, the number of serotonin types grew to five, and by 2000, seven different serotonin "families" and at least fifteen different subpopulations had been identified. The serotonins are related in that they are made of a base molecule of tryptophan (a simple amino acid) and they all have an oxygen-hydrogen group in the fifth position of the base molecule. Beyond that similarity, serotonins vary widely in chemical composition, location of receptors, and the actions they cause. Trying to track exactly what version of serotonin affects which receptors and precisely how they do so, "has become a nightmare for those involved in identifying or developing site-selective [drugs]" according to experienced drug developers.[17]

SSRIs

The "SSRI" moniker was a mere marketing concept based on an unproven hypothesis with little evidence, but it was used in a very effective campaign to maximize acceptance of the drugs. Inventing the hypothesis that mental distress is strictly a biochemical problem and knowing the drugs affect brain chemistry, marketing departments fashioned spin to claim that the drugs balanced that brain chemistry. This was the twentieth-century version of the old mystical balancing of humors. The SSRI myth quickly turned into presumed fact. Scientists have looked in vain for any consistent data to support the notion of a serotonin deficiency in depressed people.

Serotonin levels are not measured in patients for two reasons: First, the level of serotonins in the bloodstream bears no relationship to serotonin levels in the brain. Second, the brain level of serotonins is constantly fluctuating and cannot be directly measured.

---

17 R. Glennon et al., "Serotonin Receptor Subtypes and Ligands," Neuropsychopharmacology, American College of Neuropsychopharmacology, 2000, acnp.org/g4/GN401000039/Ch039.html.

There have been hundreds of studies trying to prove the serotonin hypothesis, but none have provided the conclusive evidence drugmakers need. An experiment intentionally lowering the serotonin levels of healthy subjects failed to cause depression.[18] Opposite experiments were also carried out; mood did not improve when huge increases in serotonin levels were caused by direct injection of serotonin into depressed subjects.[19] After a comprehensive review of the extensive medical literature on serotonin, the German medical board concluded reports of serotonin deficiency supposedly being responsible for violent suicide attempts were based on flawed scientific methods.[20]

In the US, drug companies are supposed to compile research results and provide the data to the FDA to validate that their drugs are effective. A close look shows that in about half of such studies submitted between 1987 and 1999, the drugs were no more effective in treating depression than placebo. It turns out the studies unfavorable to their drugs were not sent in for publication in medical journals, so they stayed largely out of view to physicians.

Axelrod stayed at the NIMH through retirement and continued researching as professor emeritus. He is remembered primarily for his research on brain biochemistry, but since many of the same neurotransmitters are also functional in the heart, the work he performed had benefits to the development of heart medication as well.

Julius Axelrod died in 2004 at the age of ninety-two.

---

18 G. Heninger et al., "The Revised Monoamine Theory of Depression: A Modulatory Role for Monoamines, Based on New Findings from Monoamine Depletion Experiments in Humans," *Pharmacopsychiatry* 29 (1996): 2–11.

19 J. Mendels et al., "Amine Precursors and Depression," *Archives of General Psychiatry* 32 (1975): 22–30.

20 J. Roggenbach, B. Müller-Oerlinghausen, and L. Franke, "Suicidality, Impulsivity, and Aggression: Is There a Link to 5HIAA Concentration in the Cerebrospinal Fluid?," *Psychiatry Research* 113 (2002): 193–206.

21

The Midwestern Genius

Earl Wilbur Sutherland Jr. was born in 1915 in Kansas. As a youth, Sutherland worked in his father's dry goods store. He attended college and then medical school, and spent the majority of his career in the Midwest as a creative, hardworking biochemist.

When he was a second-year medical student at Washington University in St. Louis, Missouri, Sutherland had the good luck to take a pharmacology course taught by Carl Cori, a cowinner along with his wife Gerty Cori of the 1947 Nobel Prize.[1] Cori noticed Sutherland's academic excellence and recruited him into the research lab. Sutherland's first two papers concerned the enzymes involved in the metabolism of glucose.[2] From that point he was hooked on research.

After Sutherland graduated medical school in 1942, he served in the army for two years then returned to join the faculty at Washington University and continue his biochemistry research alongside the Coris. Sutherland had the opportunity to interact with visiting scientists for months at a time, including Severo Ochoa and Arthur Kornberg, who

---

1 Gerty and Carl Cori shared half of the prize in 1947, and the other half went to Bernardo Houssay of Argentina. See *Boneheads & Brainiacs*, 187–91.

2 E. W. Sutherland, S. P. Colowick, and C. F. Cori, "The Enzymatic Conversion of Glucose-6-Phosphate to Glycogen," *Journal of Biological Chemistry* 140: 309–10; E. W. Sutherland and S. P. Colowick, "Polysaccharide Synthesis from Glucose by Means of Purified Enzymes," *Journal of Biological Chemistry* 144: 423–37.

would share the Nobel Prize in 1959. He also worked with research physician Edwin Krebs, who would be a Nobel winner in 1992.

Sutherland's initial research concerned the enzymes involved in the breakdown of the storage form of glycogen in the liver. He was especially interested in how this process was stimulated by the hormones adrenaline (epinephrine) and glucagon. He found that glucagon and adrenaline could not pass through the plasma membrane of cells, so he looked for a mystery substance that facilitated these hormones. Sutherland continued this line of investigation when he moved to Western Reserve University (now Case Western Reserve University) in Cleveland, Ohio. There in 1956, he discovered that cyclic AMP (cAMP) is the substance that transfers the effects of hormones into the cell, while the hormones remain outside the cell.[3] Sutherland called this the "second messenger" effect.

As explained in his Nobel award ceremony speech, "This means that epinephrine never enters the cell. We may visualize the hormone as a messenger which arrives at the door of the house and there rings the bell. The messenger is not allowed to enter the house. Instead the message is given to a servant, cyclic AMP, which then carries it to the interior of the house."[4] Chemists David Lipkin and colleagues back at Washington University had simultaneously isolated the same substance, and they subsequently clarified its structure.[5]

Sutherland also discovered that cAMP is manufactured in the cell membrane. He rapidly found out that cAMP has many other functions. In his Nobel lecture, Sutherland lists thirty-six processes that are influenced by the concentration of cAMP, and since then even more functions have been discovered: It not only regulates sugar metabolism

---

3   E. W. Sutherland and T. W. Rall, "The Properties of an Adenine Ribonucleotide Produced with Cellular Particles, Atp, Mg++, and Epinephrine or Glucagon," *Journal of the American Chemical Society* 79, no. 13 (1957): 3608.

4   P. Reichard, 1971 Award Ceremony Speech, NobelPrize.org, nobelprize.org/prizes/medicine/1971/ceremony-speech/.

5   D. Lipkin, W. H. Cool, and R. Markham, "Adenosine-3': 5'-hosphoric Acid: A Proof of Structure," *Journal of the American Chemical Society* 81, no. 23 (1959): 6198—203.

but also lipid (fat) metabolism; cAMP increases the cells' uptake of amino acids so that they can make protein; it acts to increase or decrease the production and activity of various enzymes; it increases the production of steroids; it increases calcium resorption in bone; it increases hormone activity in the kidney; it increases the force of heart contractions; cAMP reduces tension in smooth muscle, such as in the walls of blood vessels, the stomach, and intestines and the bladder and womb; cAMP increases saliva production; increases stomach acid; and increases the release of insulin and thyroid hormones. When Sutherland was given his Nobel medal, the presenter said, "When you discovered cyclic AMP you discovered one of the fundamental principles involved in the regulation of essentially all life processes."[6]

Sutherland's breakthrough discovery occurred when he was about forty-one years old, certainly not conforming with Albert Einstein's assertion that "a person who has not made his great contribution to science before the age of thirty will never do so." Einstein is famous for his theory of special relativity, which he formulated at the age of twenty-six in 1905, but that is not what earned him the Nobel Prize in Physics.[7] In the series of four papers all published in 1905, Einstein also elucidated the photoelectric effect, which was the reason for his 1921 Nobel Prize.[8] However, there is some support for the notion that early work gets the gold. A 2016 study found that prizewinning papers of Nobelists tended to occur disproportionately early in the sequence of their lifetime of published papers.[9]

---

6   P. Reichard, Award Ceremony Speech, 1971, NobelPrize.org, nobelprize.org/prizes/medicine/1971/ceremony-speech/.

7   A. Einstein, "Zur Elektrodynamik bewegter Körper" *Annalen der Physik* 17, no. 10 (1905): 891–921; See also an English translation, A. Einstein, "On the Electrodynamics of Moving Bodies," in *The Principle of Relativity*, trans. G. B. Jeffery and W. Perrett (London: Methuen and Company, 1923).

8   A. Einstein, "On a Heuristic Viewpoint Concerning the Production and Transformation of Light," *Annalen der Physik* 17 (1905): 132–48.

9   Sinatra et al., "Quantifying the Evolution of Individual Scientific Impact," *Science* 354, no. 6312 (2016).

It is difficult to identify Sutherland's precise prizewinning paper, because the official statement of the Nobel organization is that Sutherland won "for his discoveries concerning the mechanisms of the action of hormones." That encompasses a body of scientific work spanning many years before and after his 1956 discovery of the role of cAMP. By the time of the Nobel award in 1971, Sutherland was at Vanderbilt University in Nashville, Tennessee, at the peak of his career. He continued to actively research and publish, producing an astounding sixty papers from 1970 until his premature death from esophageal hemorrhage in 1974 at the age of fifty-eight.

A survey of Nobel winners shows that a successful action is definitely to work with Nobel Prize winners—in Sutherland's case, two previous and three future winners.[10] This certainly reflects the quality of the science being done at Washington University as led by Carl and Gerty Cori. Carl Cori had these observations about what made Sutherland a great scientist: "What we see at work here is a sort of hunch or secret insight plus tenacity, the ability of differentiating between important and unimportant observations, absolute reliance on the accuracy of one's results and a prodigious memory—all qualities that characterize the successful bench scientist and that Sutherland possessed in large measure."[11] Carl Cori described Sutherland further:

> First and perhaps foremost he had the gift of intuition. He could set up the right experiment at the right time without necessarily knowing why. Secondly, his intuition was strong enough to generate a remarkable degree of tenacity. . . . Thirdly, he was an excellent worker in the laboratory who could recall any of the experiments he and his associates had carried out in the past. Fourthly, Sutherland

---

10 R. J. Roberts, "Ten Simple Rules to Win a Nobel Prize," *PLoS Computational Biology* 11, no. 4 (2015): e1004084.

11 C. Cori, "Earl W. Sutherland 1915–1974: A Biographical Memoir" (Washington, DC: National Academy of Sciences, 1974).

was highly original in his concepts and did not follow current lines of thought. He had ambition, a powerful drive,
and an intensity and singularity of purpose that was most
remarkable.[12]

Sutherland himself credited curiosity and inspiration. In 1970, he said,
"Let me confess here, lest I leave a false impression, that I did not have
the welfare of future generations primarily in mind when I began my
research on [hormones]. . . . Rather, my motivation was primarily
directed to satisfying my curiosity about how these hormones acted."[13]
Sutherland was an avid fisherman, and he claimed that some of his best
scientific thoughts came while on the water.

The Nobel Prizes are announced in October, while the awards
ceremony takes place in December, with three months of celebrity in
between. In his Nobel banquet speech, Sutherland remarked that he
liked receiving letters from so many young people. "One, eleven years
old wanted to know the procedure for winning such a prize. Now I
mention this response for a reason. I am fully convinced that medical
research can offer one a happy and productive life. And if one has a
little [V]iking spirit he can explore the world and people as no one else
can do. The whole medical research area is wide open for exploration
and I believe soon for productive application."[14]

12 C. Cori, "Earl W. Sutherland 1915–1974: A Biographical Memoir" (Washington, DC:
National Academy of Sciences, 1974).

13 As quoted in C. Cori, "Earl W. Sutherland 1915–1974: A Biographical Memoir"
(Washington, DC: National Academy of Sciences, 1974).

14 E. W. Sutherland, Banquet Speech, December 10, 1971, NobelPrize.org, nobelprize.
org/prizes/medicine/1971/sutherland/speech/.

22

Untangling Antibodies

Rodney Porter and Gerald Edelman shared the 1972 Nobel Prize in Medicine for the separate research they each did that revealed the chemical structure of antibodies.

Antibodies are chemical substances produced by the body that lock onto foreign substances needing attack and neutralization. These foreign substances are given the complementary name of antigens. In some cases, such as viruses, antibodies can disable antigens directly. In other cases, such as with bacterial antigens, antibodies bind to surface proteins on the bacterium's surface, thereby signaling to the rest of the immune system that it should be destroyed.

Antibodies had been recognized for decades by the time Porter and Edelman began their work. Antibodies were known to be very large substances, and conventional wisdom considered them mystery blobs that couldn't be approached by the biochemistry techniques of the day. The brilliance of these Nobel Prize winners is that they attacked the biochemistry problem without regard to the naysayers who said it simply could not be done.

Rodney Robert Porter was born near Liverpool in 1917, and he had an early fascination with chemistry. After getting his undergraduate degree in biochemistry, he joined the army and took part in military campaigns in North Africa and Italy. After the war, he earned his PhD at Cambridge, where he wrote a thesis on antibodies. Porter's Nobel Prize–winning research was done along with Elizabeth (Betty) Press, his constant research colleague, in the London labs of the National Institute for Medical Research. Press held a bachelor's of science and had hospital lab experience before joining Porter.

Porter's hypothesis was that antibodies had a Y-shaped structure. Porter and Press isolated antibodies and then subjected them to papain, a protein-digesting enzyme derived from the papaya. Papain fractured the antibody molecules such that the structure of the smaller pieces could be more easily studied. When Porter and Press broke up the large antibody molecule, they did indeed find it was composed of a base and two arms.[1] They demonstrated that the two arms of the Y interacted with antigens, while the base of the Y attached the antigen-antibody complex to immune cells to be processed. They further found that the base of the molecule was unchangeable, but the arms had variable configurations. This structure provided support for the theory of how antibodies function: they needed to fit lock and key with whatever antigen they encountered, so the arms of the antibody were shaped in just the right way to lock onto an antigen.

Betty Press stayed with Porter in his research labs when he moved to St. Mary's Hospital Medical School in London, and she moved with him again when he became chair of the Department of Biochemistry at Oxford University.[2] She was an expert in protein chemistry, publishing twenty-two papers and even supervising graduate biochemistry students despite having only the inferior degree. Press was the lead author on a paper that revealed new structural details about the variable portion of the antibody, and also found that at least two genes

---

1   J. B. Fleischman, R. R. Porter, and E. M. Press, "The Arrangement of the Peptide Chains in γ-globulin," *Biochemical Journal* 88, no. 2 (1963): 220–28.

2   N. Hogg and L. Steiner, "Elizabeth (Betty) Marian Press (1920–2008)," *Biochemist* (London) 31, no. 3 (2009): 50–51.

were involved in the synthesis of portions of the antibody.[3] Betty Press retired in 1980 and died in 2008 at the age of eighty-eight.

Porter made no mention of Betty Press in his Nobel lecture although he cited four of her papers.[4]

In 1985, after eighteen years at Oxford, Porter had planned to move to direct the immunochemistry lab at the Medical Research Council, but weeks before the anticipated move, he died in a car accident at the age of sixty-seven.

Gerald Edelman did not work with Rodney Porter, but he pursued a parallel line of antibody research. Edelman was born in 1929 in New York City, and as a youth, he thought he'd become a concert violinist. He soon learned that although he liked playing the violin, he did not have the temperament for stage performance nor a talent for orchestral composition. He was not a great student and had a hard time getting into college, which he attributed to a deprecating remark made in his letters of recommendation by one of his high school teachers. He finally managed to get accepted to Ursinus College in Pennsylvania through a family connection, but he found that he hated lectures and soon dropped out. He returned under the condition that he could access the library and laboratories on his own and would not be required to attend formal classes. Next, he went medical school at the University of Pennsylvania, though he never intended to practice medicine, having become much more interested in basic science. After a stint in the peacetime army, Edelman attended the Rockefeller Institute (later Rockefeller University) for his PhD work, where he did the laboratory experiments that won him the Nobel Prize.

----

3   E. M. Press and N. Hogg, "Comparative Study of Two Immunoglobulin G Fd-Fragments," *Nature* 223 (1969): 607–10.

4   R. Porter, "Structural Studies of Immunoglobulins," Nobel Lecture of December 12, 1972, in *Nobel Lectures, Physiology or Medicine 1971–1980*, ed. Jan Lindsten (Singapore: World Scientific Publishing, 1992).

In 1957 he successfully broke up the large molecule known as an antibody and was able to discern its component parts, finding a Y-shaped structure made up of two smaller chains and two larger chains. Over the next two years he conducted a total of 169 experiments before feeling confident enough to offer his results for publication. He then chose the most diffident way to get something in print—as a letter, not to a medical journal but to the *Journal of the American Chemical Society*.[5]

Despite Edelman's attempt to fly under the radar with his discovery, the letter was broadly noticed and sparked several lines of research into the immune system, eventually yielding great advances in the knowledge of how antibodies react with antigens. Edelman remained affiliated with Rockefeller University and continued in the field of immunology for the next two decades, with about seventy publications on which he was listed as the lead or sole author and twice that many where was listed as a coauthor. In the latter cases, it is unclear how many of the articles reflected his own work or that primarily done by the doctoral students in his labs.

Edelman's experience of the December 1972 Nobel ceremonies was influenced by the hefty dose of tincture of opium he was taking at the time for some minor bowel troubles. "I said my God, these Swedes have some hi-fi set! It happened to be the Swedish Philharmonic [playing] behind me. People said my eyes were so marvelous. I was in a complete trance."[6]

In the late 1970s, Edelman abandoned the field of antibody research to focus on the neurosciences, specifically the structure and function of the brain. Starting in 1992 until his retirement, Edelman headed the Department of Neurobiology at the Scripps Research Institute. He continued to publish frequently, contributed to textbook chapters, and

---

5   G. M. Edelman, "Dissociation of y-Globulin," *Journal of the American Chemical Society* 81 (1959): 3155.

6   G. Edelman, Interview, September 2008, NobelPrize.org, nobelprize.org/prizes/medicine/1972/edelman/interview/.

wrote books for the lay public, but his writing largely concerned the theoretical. He made no major discoveries in the neuroscience field. In a 2004 interview, Edelman remarked that over the course of his fifty-year career, "I have something like 555 papers and publications, mostly with others but sometimes not, and I suspect that about 7 of them are good."[7]

Edelman developed a theory he called neural Darwinism, which posited that the human brain is consciousness.[8] It was based on the observation that the human cerebral cortex has an estimated thirty billion neurons and the brain as whole has a million billion connections. The theory supposed that a series of lucky accidents brought about increasing nerve-to-nerve connections, each of which gave a survival advantage and thus was "selected for" in Darwinian fashion as subsequent generations reproduced. Eventually, so the theory goes, brain size and connectivity reached a critical mass that Edelman equated with consciousness. He was insistent that human brain-nerve connectivity is not the cause of consciousness but rather is itself consciousness.

The basic idea that we are our brains was not new or original, but Edelman dressed it up in language borrowing from anatomy, biology, genetics, psychology, psychoanalysis, and a variety of philosophies.[9] Francis Crick, a Nobel laureate who also dove into the nether regions of the brain and came up empty handed, thought the theory was better termed "Neural Edelmanism."[10] Edelman wrote extensively on this topic but his lengthy and tortuous descriptions and rationales are short on objective data and long on theory. By the time of his sixth and final popular book, Edelman confessed his own confusion on the subject: "Consciousness is to me a mystery, and not one to be dismissed." His

---

7   G. Edelman, Interview, September 2008, NobelPrize.org nobelprize.org/prizes/medicine/1972/edelman/interview/.

8   G. M. Edelman, *Neural Darwinism: The Theory of Neuronal Group Selection* (New York: Basic Books, 1987).

9   G. Edelman and G. Tononi, *A Universe of Consciousness: How Matter Becomes Imagination* (New York: Basic Books, 2000).

10 F. Crick, "Neural Edelmanism," *Trends in Neurosciences* 12, no. 7 (1989): 240–48.

writing does little to explain much less convince one of the basis for his conclusions, and he admitted this shortfall in communication: "We know what it is like to be conscious, but not how to put it into satisfactory scientific terms." Yet in the very next sentence he asserted, "Whatever it precisely may be, consciousness is a state of the body, a state of nerves," which is contradicted by his concession that "we have not yet scientifically founded the basis of consciousness in the brain." Ultimately, Edelman said, "We must admit that our knowledge of knowledge has great gaps."[11]

It is a profound tragedy that Edelman, who considered himself to be his brain, suffered from Parkinson's disease, a degenerative brain disorder. Signs of the disease are evident in the 2004 video interview of Edelman on the Nobel website, in which he rarely blinks and his face lacks animation (a disease characteristic called "masked facies").[12] Gerald Edelman died in 2014 at the age of eighty-four.

11 G. M. Edelman, *Second Nature: Brain Science and Human Knowledge* (New Haven, CT: Yale University Press, 2006).

12 G. Edelman, Interview, September 2008, NobelPrize.org, nobelprize.org/prizes/medicine/1972/edelman/interview/.

23

The Birds and the Bees

In 1973, the Nobel Prize in Physiology or Medicine was awarded jointly to Karl von Frisch, Konrad Lorenz, and Nikolaas Tinbergen for their observations of animal behaviors.

The story of Karl von Frisch's research provides a refreshingly enjoyable Nobel tale. He won the prize for his simple but brilliant experiments with bees, revealing an astonishingly complex array of behaviors that causes one to immediately rethink the concept that clever adaptability is in any way connected to brain size.

Von Frisch was born in 1886 in Vienna, and his childhood bedroom contained 123 animal species: "9 species of mammals, 16 species of birds, 26 cold blooded vertebrates, 27 fish and 45 invertebrates."[1] He delighted in observations about animals, noticing that his fish responded to a whistle for food although they had no obvious ears. Von Frisch subjected this to rigorous experimentation thirty-five years later, and it resulted in a publication in a scientific journal—the conclusion: Fish can hear![2] Both his father and grandfather were physi-

1   E. Etxebeste Aduriz, "Karl von Frisch: Dancing with Bees," *Elhuyar Science*, November 30, 2018.
2   K. von Frisch and R. Stetter, "Untersuchung über den Sitz des Gehörsinnes bei der Elritze" [Examination into the position of the sense of hearing in the minnow], *Zeitschrift für vergleichende Physiologie* 17 (1932): 686–801.

cians, and initially von Frisch followed the family tradition by attending medical school. After his first exams, he abandoned medicine for the study of zoology, in which he earned a doctorate.

His first bee experiment sought to challenge the widely held belief that bees and other insects were color-blind. That just did not make sense to von Frisch, given the variety of brightly colored flowers regularly visited by pollinators. The experiment was simple. He set up a table outdoors and placed on it a colored paper between other papers that were shades of gray. He put a small dish of syrup on the colored paper. Scout bees located the sugar readily, and in a few minutes the dish was found by a horde of forager bees. Then he replaced the syrup with an empty dish. During the break in feeding, a few scout bees still visited the empty dish on the colored papers—but not the gray papers. When he again placed a syrup dish on the colored paper, the entire forager group of bees returned to it.[3]

These observations made von Frisch question how the scout bees communicated the location of the feeding place to the forager bees. He tracked the activity of bees feeding on robinia flowers and a different group of bees (from the same hive) feeding on linden blossoms. All of the bees mingled during a feeding break. Afterward, when scouts returned from the robinia flowers, only the robinia foragers paid attention the dance of the robinia scout, who were ignored by the linden foragers. The robinia foragers then flew back out to the robinias. When the linden scout returned, only the linden foragers paid attention to the dance of the linden scout, and the linden foragers flew out to feed again on linden. There were only a few exceptions when individual enterprising bees would switch sides and take advantage of both kinds of flowers.

As in all good experiments, this prompted further inquiry: Do bees communicate by smell? Near the hive, von Frisch placed a dish of syrup on a piece of cardboard scented with peppermint. After the

3   K. von Frisch, "Der Farben und Formensinn der Bienen" [The bee's sense of color and shape], *Zoologische Jarbücher (Physiologie)* 35 (1914–15): 1–188.

bees had a good feeding, von Frisch removed the nearby dish. Next, he placed dishes of syrup on cards in various locations out in the field, some of them scented with peppermint and some unscented. The bees invariably went to the dishes on the scented cards and ignored the others.[4] The findings led to an announcement in the *New York Times*: Bees can smell![5]

In this experiment, von Frisch noticed there was an additional element to the communication in the form of the dance of the bees: "If the sugar syrup becomes scarce or is offered in weaker concentrations, after a certain point the dancing becomes slower and finally stops even though the collecting may continue. On the other hand, the sweeter the sugar syrup, the more lively and lengthier the various dances."[6] Von Frisch next demonstrated this was true in the natural setting: "Various types of plants blossom simultaneously, producing nectar of differing concentrations. The richer and sweeter its flow, the livelier the dance of the bees that discover and visit one type of flower. The flowers with the best nectar transmit a specific fragrance which ensures that they are most sought after."[7]

---

4   K. von Frisch, "Über den Geruchssinn der Bienen und seine blütenbiologische Bedeutung" [The bee's sense of smell and its significance during blooming], *Zoologische Jahrbücher (Physiologie)* 37 (1919): 1–238.

5   "Sense Of Smell in the Bee Tested by German Scientist; Professor von Frisch Believes That It Guides the Insect to Various Flowers," *New York Times*, April 3, 1927 (consulted from archives).

6   K. von Frisch, "Decoding the Language of the Bee," December 12, 1973, NobelPrize. org, nobelprize.org/uploads/2018/06/frisch lecture.pdf. As Karl von Frisch was unable to be present at the occasion, the Nobel Lecture was delivered by his son Professor Otto von Frisch.

7   K. von Frisch, "Decoding the Language of the Bee," December 12, 1973, NobelPrize. org, nobelprize.org/uploads/2018/06/frisch-lecture.pdf. Other bee researchers found the same characteristics of dances conveyed information on the quality of a potential location of a new hive. The "swarm" is when separate groups of scouting bees return with messages conveyed in dances. The scouting team displaying the most intense dance wins as showing the best domicile. There is nothing artistic about it, the dance simply being a description of location along with intensity communicating an unembellished report of the quality of the site. See, M. Lindauer, "Swarm Bees Looking for Accommodation", *Magazine for Comparative Physiology* 37, no. 4 (1955): 263–324.

Such was the state of von Frisch's knowledge in the 1920s, by which time he figured he pretty much knew all there was to know about bees. Von Frisch served in teaching and research capacities at a couple of German universities before settling at the University of Munich in a state-of-the-art zoology institute funded by the Rockefeller Foundation.

The Nazi Party gained political control in Germany, and by 1938, they had annexed Austria. Von Frisch eventually received anonymous warnings that he was allowing too many Jews to work in his lab; the Law for the Restoration of the Professional Civil Service of April 1933 excluded Jews and the "politically unreliable" from civil service. The number of Jewish students at German schools and universities was restricted, and further legislation sharply curtailed Jewish participation in the medical professions. This anti-Semitism was made even more harsh by the 1935 Nuremberg Race Laws, which defined Jews by ancestral lineage rather than by religious practice or belief, made Jews noncitizens of Germany, and required physical segregation of Jews.

Von Frisch had enemies at the zoology institute, including botanist Ernst Bergdolt, who was the president of the University National Socialist League, and Wilhelm Führer, who led the university's Instructor's League. These two enthusiastically participated in *Gleichschaltung*, the alignment of all academic, social, political, and cultural organizations according to Nazi ideology and policy. This involved Nazification of every aspect of life, for which prominent academics at elite universities and research institutes demonstrated their support by providing scientific-sounding rationale to legitimize racial-eugenic policies and the führer style of leadership. Bergdolt reported to the Ministry of Education that von Frisch failed to promote his bee research as aligning with Nazi ideology.[8] Von Frisch refused to extrapolate from the bee society to human society, although it would have been easy for him to state that beehive order, conformity,

---

8   S. Najafi and H. Raffles, "The Language of the Zoologische Jarbücher," *Physiologie* 35 (1914–1915): 1–188; S. Najafi and H. Raffles, "Bees: An Interview with Hugh Raffles; Karl von Frisch and His "Little Comrades," Cabinet Magazine, Spring 2007.

hierarchy, and leadership was nature's example of the rightness of the Nazi state. (In contrast to von Frisch, Konrad Lorenz, his Nobel Prize cowinner, was a great asset to Nazi propaganda when he made scientific-sounding arguments about the relationship between race and animal instinct.) It was dangerous to voice opposition to the Nazi regime, but von Frisch was not inactive otherwise. His biographer recounts von Frisch's role in helping to release a Polish scientist from the Dachau concentration camp in 1940.[9]

In early 1941, von Frisch received a letter from the government: "We force you to leave, according to the law of 1937, because you do not meet the requirements to be a teacher."[10] This was perhaps in reference to a 1938 Reich supreme court declaration that being a Jew was cause for dismissal from a job. Von Frisch's heritage had been investigated, and it was discovered that his maternal great-grandparents were Jewish converts to Catholicism. The letter declared that he was a "second-degree crossbreed."[11]

The bees saved him. In the war years, there was an estimated loss of eight hundred thousand hives across Europe caused by disease from the parasitic Nosema fungus. The resultant lack of pollinators on food crops made wartime food scarcity even worse. A special commission was set up to make recommendations on the bee catastrophe—and von Frisch's future, presumably a choice between staying on in service to the Reich or potential murder in a concentration camp. One of von Frisch's strongest supporters was the president of the Association of South Bavarian Beekeepers, who wrote to Nazi headquarters describing von Frisch as "the most successful bee researcher of the world" and imploring them to spare the scientist in order to help resolve the bee

9   T. Munz, *The Dancing Bees: Karl von Frisch and the Discovery of the Honeybee Language* (Chicago: University of Chicago Press, 2016).

10 As quoted in E. Etxebeste Aduriz, "Karl von Frisch: Dancing with Bees" *Elhuyar Science* November 30, 2018.

11 T. Munz, *The Dancing Bees: Karl von Frisch and the Discovery of the Honeybee Language* (Chicago: University of Chicago Press, 2016).

disaster.[12] Von Frisch was allowed to stay at the zoology institute to work on the Nosema plague.

Nosema Disease in Honeybees

Nosema is a single-celled fungus that lives and reproduces in the honeybee gut, where it steals nutrients and prevents digestion. The organism has a spring-loaded lancet that injects infectious spores (dormant forms of the parasite) into the cells lining the bee intestine. There, the spores become animated and reproduce, inhibiting the formation of digestive enzymes. Instead of the cells erupting to release digestive enzymes, they release mature Nosema, each with its own spore-injecting lancet, which go forth to inject other cells. This causes the bee to starve despite having plenty of food. Nosema infection reduces a worker bee's lifespan by 50 to 75 percent, and it is thought to be the cause of colony collapse disorder, the sudden dwindling of a colony in the early spring.

There are eighty-two known species of Nosema, with specific species selectively infecting mosquitoes, moths, beetles, and bees. Historically Nosema in honeybees was caused by *Nosema apis*, but in 2007, another Nosema species was detected in honeybees, *Nosema cerana*, from Asia. It seems to strike later in the spring and into early summer. The new Nosema caused widespread colony collapse initially, but then things leveled off. This is explained by the well-known phenomenon that holds true across most species, including humans: when a new disease-causing organism arrives on the scene of a previously unexposed population, the first wave of disease is usually the worst because the most susceptible organisms quickly become infected. Later, some immunity develops, causing the prevalence of the disease to decrease. In 2017, another new Nosema, *Nosema neumanni*, was found in honeybees in Africa.

---

12 T. Munz, *The Dancing Bees: Karl von Frisch and the Discovery of the Honeybee Language* (Chicago: University of Chicago Press, 2016).

Almost all hives harbor some Nosema, and it can even be found in high concentrations in apparently healthy, thriving hives. Therefore Nosema is considered to be an opportunistic disease, meaning that it will overgrow and cause illness only in the most vulnerable when they are stressed by other factors, such as pesticides, malnutrition from feeding on genetically modified plants, or a simultaneous infection.

There are no safe antibiotics that treat Nosema, but one essential oil from a Chilean evergreen tree, *Cryptocarya alba*, has been found to be effective.[13] The best way to avoid illness is by prevention—making sure the colonies have good foraging options and are not exposed to pesticides.

Von Frisch managed to hold on to his job, although it is doubtful if he was able to do anything effective to combat Nosema. Three years later, over the course of one day and one night in April 1944, the city of Munich was carpet-bombed by US and English air forces, dropping a total of 450 large bombs, 61,000 high-explosive bombs, and 3,316,000 magnesium incendiary bombs. There were 6,632 people killed, 15,800 wounded, and more than 300,000 left homeless. It also destroyed the zoological institute, but von Frisch was unharmed.

After the war, von Frisch relocated to Graz in Austria and returned to his studies on how bees communicate, publishing amazing observations. He found that when scout bees returned from a nearby feeding place, they danced in circles (the round dance), but when they returned from feeding sites more than fifty meters away, they danced with a tail-wagging motion (the tail-wagging dance). This information alone was not too useful, but he found the dancing was combined with three additional data relays that specified just how far the feeding location was: 1.) the pace of the tail wagging decreases with increasing distance; 2.) the dance is simultaneously toned with a buzzing sound;

13 V. V. Ebani and F. Mancianti, "Use of Essential Oils in Veterinary Medicine to Combat Bacterial and Fungal Infections," *Veterinary Sciences* 7, no. 4 (2020): 193.

3.) longer distances are expressed by longer tail-wagging times.[14] Von Frisch was able to precisely plot these dances against the distances they represented.

These observations were made on the *Apis mellifica carnica* species. Other researchers have found that the distance at which scout bees transition from round dance to tail-wagging dance is thirty meters for *A. mellifica mellifica* and *A. mellifica intermissa*, about twenty meters for *A. mellifica caucasia* and *A. mellifica ligustica*, and seven meters for *A. mellifica fasciala*. To his embarrassment, von Frisch realized he had observed these different dance styles some twenty years before, but mistakenly thought the tail-wagging dance was simply characteristic of pollen-basket carriers as opposed to sugar-syrup collectors—he had missed the fact that the pollen-basket carriers always came from farther afield.

Von Frisch's most intriguing discoveries regarded the bees' perception and communication of direction. He noticed that when the scout bees danced, they always oriented their bodies in the direction of the feeding site. In advanced bee species, this body orientation is maintained even inside a dark hive on a dance floor that is a vertical wall of the hive. No matter how the hive is turned, the scouts correctly orient their bodies like a needle in a compass.

Von Frisch discovered bees use the position of the sun as a compass, and can do so even when the sun is occluded by clouds or a mountain.[15] Bees see polarized light, something humans cannot do naturally. When humans look at the sky, they see blue or maybe cloud cover; bees perceive lines of polarized light in the sky. The dancing scout aligns its body with the lines to convey the direction to the feeding site "as the crow flies," no matter if the scout actually took detours

---

14 K. von Frisch, "Die Tänze der Bienen" [The bees' dances], *Österreichische Zoologische Zeitschrift* 1 (1946): 1–48.

15 K. von Frisch, "Die Sonne als Kompaß im Leben der Bienen" [The sun as compass in the life of bees], *Experientia (Basel)* 6 (1950): 210–21.

on his travels due to terrain and obstructing objects.[16] It has since been shown that other animals, such as ants, crayfish, spiders, and octopuses use polarized light for their individual orientation, but only bees additionally use it to communicate the direction to a distant site. Subsequent research has found that bees also use gravity, ultraviolet light, and an internal clock to pinpoint directions and distances.

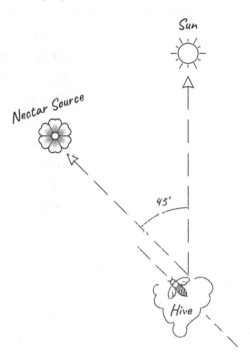

Von Frisch returned to Munich in 1950, when the zoological institute was rebuilt, and continued research after retirement as emeritus professor. His Nobel Prize was awarded when he was eighty-seven years old, fifty-nine years after his initial bee research. He could not attend the ceremony and was represented by his son, who read his lecture. All insect researchers who have followed owe a debt to the pioneering work of Karl von Frisch. He died in 1982 at the age of ninety-five.

---

16 K. von Frisch, "Die Polarisation des Himmelslichtes als orientierender Faktor bei den Tänzen der Bienen" [The polarisation of skylight as a means of orientation during the bees' dances], *Experientia (Basel)* 5 (1949): 142–48.

Unlike Karl von Frisch, Nobel cowinners Konrad Lorenz and Nikolaas Tinbergen readily translated their observations of animal behavior into significance for the human realm.

Konrad Lorenz was born in Vienna in 1903, when biology was attaining dominance as the scientific basis for all things natural. Lorenz first became interested in ethology (the study of animal behavior) as a youth, when a gosling followed the boy around on the grounds of his family estate and eventually adopted the young Konrad as his parent. The human boy followed in his own father's footsteps by attending medical school before working in comparative anatomy at the University of Vienna. When the head of the department retired, Lorenz's enthusiasm for drawing correlates of animal behavior to human motivations was not appreciated by his new superiors. He was banned from ethologic work and no longer welcome on academic staffs in Austria, and so he pursued animal studies privately with the support of his physician-wife's income. He was later awarded an honorary doctorate in zoology.

In 1936, Konrad Lorenz met his future cowinner Niko Tinbergen when he visited Leiden University in Holland for a symposium on instinct. That summer the Tinbergen family were house guests of Lorenz for four months, and the scientists collaborated on their animal observations. Also in 1936, Lorenz received his habilitation, meaning that his academic and personal views were sufficiently pro-Nazi to allow his academic employment. However, his scholarship application to the German Research Foundation was rejected in 1937 on the grounds that "the political convictions and descent of Dr. Konrad Lorenz [are] questioned"—probably because his wife's sister was married to a Jewish industrialist.[17] On his handwritten application for membership in the National Socialist German Worker's [Nazi] party

---

17 Attributed to Fritz von Wettstein in a statement dated December 14, 1937, as quoted in U. Deichmann, *Biologists under Hitler: Displacement, Careers, Research* (Frankfurt am Main, 1992), 250.

the next year, he attempted to erase any lingering doubt about his fitness as a Nazi:

> As a German thinker and scientist, I was of course always
> a National Socialist and, for ideological reasons, a bitter
> enemy of the black regime (never donated or flagged)
> and because of this attitude, which also emerged from my
> work, had difficulties obtaining the lectureship.[18] I developed a really successful advertising activity among academics and especially students, long before the upheaval
> I had succeeded in proving the biological impossibility
> of Marxism to socialist students and converting them to
> National Socialism. On my many congress and lecture
> trips, I always and everywhere tried with all my might to
> counter the lies of the Jewish international press about the
> alleged popularity of Schuschnigg and about the alleged
> rape of Austria by National Socialism with compelling
> evidence.[19] I did the same to all foreign work guests at my
> research center in Altenburg. Finally, I can say that all of
> my life's scientific work, in which questions of genealogy,
> race and social psychology are in the foreground, is in the
> service of National Socialist thought![20]

Lorenz was granted Nazi Party membership, and he also served in the Nazi Office of Racial Policy. Lorenz wrote and spoke enthusiastically on the need for eliminating the ethnically inferior elements from human society, as in this quote from his presentation at the Sixteenth Congress of the German Psychological Association in 1938: "Nothing is more important for the health of an entire people than the elimination of

---

18 "Black regime" is a possible reference to the black eagle on the Austrian coat of arms and flag.

19 A reference to Kurt Alois Josef Johann Schuschnigg, who was the chancellor of the Federal State of Austria from 1934 until the 1938 anschluss with Nazi Germany.

20 K. Taschwer and B. Föger, *Konrad Lorenz: Biography* (Vienna: Zsolnay, 2003): 84–85.

invirent [weak] types, which, with the most dangerous and extreme virulence, threaten to penetrate the body of a people like the cells of a malignant tumor."[21] In 1940, he warned that even the current efforts at racial cleansing might not be rigorous enough.

Invirent

Invirent is a word coined by Nazi psychologist Erich Rudolf Jaensch to mean "weak" or "disintegrative" yet also highly threatening to the health of a people. Jaensch, a contemporary of Lorenz, proposed a biological basis for distinctions between two personality types. The J type would make a good Nazi: J made definite, unambiguous perceptual judgments and persisted in them, recognizing that human behavior is fixed by blood, soil, and national tradition. The J type would be tough, masculine, firm—a man you could rely on. The S type was someone of racially mixed heredity and included Jews, Parisians, East Asians, and communists, characterized by what Jaensch called "perceptual slovenliness"—the qualities of one sense carelessly mixed with those of another, causing ambiguous and indefinite judgments and a lack of perseverance.[22]

In his bird research, Lorenz observed the differences in wild and domesticated geese. He noticed the domestic breeds were more driven to feed and to copulate but had less differentiated social instincts than their wild cousins. Lorenz thought he observed the same thing in human society, and he asserted that he was frightened by such social

---

21 From Lorenz's speech at the Sixteenth Congress of the German Psychological Association in 1938, as quoted in T. J. Kalikow, "Konrad Lorenz's 'Brown Past': A Reply to Alec Nisbett," *Journal of the History of the Behavioral Sciences*, 14 (1978): 173–80.

22 E. R. Jaensch, "Grundsätze fur Auslese, Intelligenzprüfung und ihre praktische Verwirklichung" [Principles for selection, intelligence measurement and its application], *Zeitschrift für angewandte Psychologie und Charakterkunde* 55 (1938): 1–14.

degeneration, considering it the genetic deterioration of humanity. He wrote about the "dangers of domestication" in a landmark 1940 article that was nothing less than an overt endorsement of racial cleansing in order to eliminate those with supposed inferior genetics from the breeding pool.[23] This pseudoscientific reasoning served to endorse racial murder, political murder, and medical murder, of which Lorenz later claimed he had no inkling. This and his other publications on the need for Nazi-style eugenics were considered a sufficiently large contribution toward the Nazi cause, and later that same year Lorenz was appointed professor as the chair for psychology at the Philosophical Faculty of the University of Königsberg.

Sketch by Konrad Lorenz from a letter to Oskar Heinroth, January 18, 1939.

---

23 K. Lorenz, "Disorders of Intrinsic Behavior Caused by Domestication," *Journal for Applied Psychology and Character Studies* 59, nos. 1–2 (1938): 2–81.

Nazi Doctors

Of all the professions, doctors had the highest ratio of Nazi Party membership, with estimates that 50–65 percent of German doctors joined the party. This was entirely voluntary, according to fifty years of postwar research summarized by H. Friedlander, who wrote, "Proof has not been provided in a single case that someone who refused to participate in killing operations was shot, incarcerated, or penalised in any way."[24]

Even more chilling is that physicians joined the Schutzstaffel (SS) seven times more than teachers or average male employees.[25] The SS was initially a small group of Hitler's bodyguards, but under Heinrich Himmler, it eventually expanded to over 250,000 members (some estimates put this number near one million). The SS became the most feared Nazi division, and a subdivision of the most ruthless, the Waffen SS, comprised the staff that ran the concentration camps and conducted brutal pseudomedical experiments.

While they did not invent eugenics, Nazi doctors built up the scientific-sounding rationale around it, contributed to the laws, wrote the criteria for selection of the unfit, and even handpicked victims for sterilization and murder. The doctors most valuable to the Nazi cause and most enthusiastically participating were the psychiatrists, psychologists, and social biologists—who have been identified as the worst transgressors of all. It was actually psychiatrists who initiated and carried out the organization of mass extermination, providing the driving force of Nazi eugenics. "Their role was central and critical to the success of Nazi policy, plans, and principles."[26]

24 H. Friedlander, *The Origins of Nazi Genocide: From Euthanasia to the Final Solution* (Chapel Hill: The University of North Carolina Press, 1995), 235–6.

25 V. Roelcke, "Academic Medicine During the Nazi Period and Implications for Creating Awareness of Professional Responsibility Today" in S. Rubenfeld, ed., *Medicine after the Holocaust* (New York: Palgrave Macmillan, 2010).

26 R. D. Strous, "Psychiatry during the Nazi Era: Ethical Lessons for the Modern Professional," *Annals of General Psychiatry* 6(2007):8, doi.org/10.1186/1744-859X-6-8.

In 1941, Lorenz was drafted as an army psychiatrist and assigned to German-occupied Poland. There he volunteered his services in a "study" known as the Polish Selection that involved 877 people. German Poles from families who had immigrated from eastern German lands were selected for relocation in the Polish region annexed by Germany. The remaining locals underwent a psychological evaluation to determine whether they were "biologically inferior," in which case they were sent to concentration camps.[27] Lorenz's later denial of knowledge of the murders he was facilitating is unbelievable.

Lorenz was sent on a German mission to the Soviet Union in April 1944, and he was soon taken prisoner but allowed to work as a prison camp doctor. Lorenz did not describe undue hardships, and he boasted that he got along well with the Russians, especially their doctors. He was released in 1948.

Lorenz proved himself to be excellent at ingratiation, first as a Nazi, then as a prisoner of war in the Soviet Union, and again when he returned home. He realized he would have better chances with colleagues in Germany than in Austria, where former Nazis were no longer welcome. In 1949, the Max Planck Institute set up the Research Center for Comparative Behavioral Research specifically for Lorenz. He relied on connections to fellow racists in the British Eugenics Society to establish friendly relations at Oxford. Lorenz even mended relations with his eventual Nobel cowinner Niko Tinbergen, who had been in a Nazi prison camp in occupied Holland during the war.

Lorenz has been esteemed as "a founding father of ethology,"[28] which ignores the fact that the word was coined by John Stuart Mill

27 R. Hippius et al., "Volkstum, Gesinnung und Character" [Reports on psychological examinations of German-Polish mongrels and Poles from Poznan] Summer 1942. Stuttgart/Prague, Verlag W. Kohhammer, 1943. Lorenz is mentioned as a collaborator in the introduction to the book and had a copy with Hippius's handwritten dedication, "Meinem Forschungskameraden" [To my fellow researcher].

28 See, for example, "A Talk with Konrad Lorenz," *New York Times Magazine*, July 5, 1970.

in 1843.[29] For two centuries before that, the people studying animal behavior had been called zoologists or simply naturalists. Lorenz is famous for his observations that some animals will follow the first animate object encountered after leaving the nest. This behavior was noticed in geese by Lorenz when he was a child, and indeed had been noticed by others for centuries if not millennia. His main contribution to the field was to name it "imprinting."

Lorenz has been referred to as the Einstein of ethology, and, like Einstein, he was awarded the Nobel Prize primarily for his theories rather than for actual discoveries. Prime among these were his theories of what drives behaviors, for which he categorized some activities as innate (instinctual). These he presumed to be hardwired in the brain and determined by genes. Other activities, he said, were not innate but undertaken in order to lead up to the instinctual act. Lorenz explained the nuances of these actions by an elaborate theory of "inducers," those things in the environment that would trigger an instinct into action via some brain wiring, and "repressors," a supposed brain mechanism that held the instinct in check until it was sufficiently stimulated to overcome the suppression. A typical example of animal observations that Lorenz explained by his theories is this description of a cat: "If you feed that cat a great number of mice, you will find that after repeating the sequence four of five times it will still kill a certain number of mice. And then, surprisingly enough, it will go on stalking mice at a distance while other mice are running over its paws. And it will not kill any more, but it will go on stalking for a long time . . . if the cat's hunger is satisfied, the drives of jumping, killing, and stalking are still not extinguished."[30] One would need to read his volumes on environmental stimuli, brain repressors, the waning of instinctual behaviors, and the so-called releasing thresholds of triggers in order to fit the cat's activity

---

29 T. Ball, "The Formation of Character: Mill's 'Ethology' Reconsidered," *Polity* 33, no. 1 (Autumn 2000): 25–48.

30 R. E. Evans and K. Lorenz, *Konrad Lorenz: The Man and His Ideas* (New York: Harcourt, 1975).

into the Lorenz way of thinking. Critics have pointed out that Lorenz's theories are not based so much on observation as a preconception that behavior falls into the two categories of innate and noninnate, and therefore all observations are crammed into that theory.[31]

Lorenz theorized on the relevance of animal behavior to human-kind, and pronounced that the need for social acceptance was rooted in human brain biology, hardwired into the central nervous system. This was closely related to his theories of aggression, which Lorenz considered to be inherent and unprovoked, with animalistic urges toward aggression only curbed in mankind by laws and customs. These theories yielded his most popular book, *On Aggression*, published in 1966, coinciding with pronounced racial unrest in the US and South Africa. Lorenz expounded in particular on youth rebellion, which he also thought was hardwired and had no relationship whatsoever to the rational reasons given for various activities such as Vietnam protests or rejection of social injustices. He saw these activities of youth as one big mental illness: "An entirely unconscious and deeply infantile revolt against all parental precepts in general, right down to those of early toilet training." This, he said, caused a war between two successive gen-erations. "The causes of this war, from the viewpoint of the ethologist as well as from that of the psychiatrist, appear to lie in a mass neurosis of the worst kind."[32] The cure, according to Lorenz, would be the teach-ing of biology, for only then the kids would understand that it was biology that was making them hate. These explanations fit nicely into the growing biological-psychiatry movement, which categorized every thought, mood, and emotion as emerging from chemical and electrical phenomena in the body.

Writing his autobiography for the blurb in the Nobel archives in 1973, Lorenz weakly apologized for the pro-Nazi terminology he

---

31 D. S. Lehrman, "A Critique of Konrad Lorenz's Theory of Instinctive Behavior," *Quarterly Review of Biology* 28, 4.

32 R. E. Evans and K. Lorenz, *Konrad Lorenz: The Man and His Ideas* (New York: Harcourt, 1975).

used in the 1940 paper while still defending his past support of the Nazi regime, claiming that at the time he was surrounded by many refined gentlemen who were also hoping the Nazis would be good for the world, including "my own father, who certainly was a kindly and humane man." The only thing Lorenz regretted was that his Nazi sympathies hampered "the future recognition of the dangers of domestication."[33] On the heels of winning the Nobel Prize, Lorenz wrote a book for the lay public called *Civilized Man's Eight Deadly Sins*, which includes a chapter on "genetic decay" that reiterates the threat of the genetically inferior.[34] In 1975 Lorenz was asked again about this Nazi involvement, but he declined to answer, saying, "If I do [answer], it will appear as if I feel the need to apologize, which I do not."[35]

Nobel award ceremonies include a royal banquet, at which each laureate is invited to read a short speech. These are usually very brief and characterized by appreciation and deep humility. Not so for Lorenz, who basically congratulated himself on the prize and how brilliant it was that he was recognized by the Nobel Committee.

Scientists all over the world tend to agree with the judgments of the expert groups that meet in Stockholm with a unanimity that is rarely found anywhere else. I have often heard differences of opinion as to whether the award of a literary Nobel Prize or a Nobel Peace Prize is justified— as is well known, one can disagree on artistic and ethical issues—but I have never attended a discussion between scientists who disagreed about whether a colleague is worthy of the high distinction or not. This worldwide

33 K. Lorenz, Biographical, NobelPrize.org, circa 1973, nobelprize.org/prizes/medicine/1973/lorenz/biographical/.

34 K. Lorenz, *Civilized Man's Eight Deadly Sins* (New York: Harcourt Brace Jovanovich, 1974).

35 R. E. Evans and K. Lorenz, *Konrad Lorenz: The Man and His Ideas* (New York: Harcourt, 1975).

unanimity of scientific opinion contains the highest recognition that can be given to the scientific staff of the
Nobel Foundation.[36]

In fact, there was considerable controversy. A headline in the *New
York Times* read, "Questions Raised on Lorenz's Prize."[37] The German
American psychiatrist Dr. Fredric Wertham said, "To give the Nobel
Prize to Lorenz does unspeakable harm, to my mind."[38]

Konrad Lorenz died in 1989. In 2015, the Associated Press reported
that Lorenz was posthumously stripped of his honorary degree: "The
University of Salzburg announced that Konrad Lorenz, a zoologist
who was awarded the 1973 Nobel Prize in Physiology or Medicine,
no longer holds his honorary doctorate, 26 years after his death." The
reason cited was his fervent participation in the Nazi efforts.[39]

———

Nikolaas Tinbergen was born in 1907 in the Netherlands. By his own
account, he was a mediocre student from the time of his earliest school
years through to university, having much more of an interest in wild
animals in the great outdoors. At Leiden University he was gratified to
interact with world class naturalists, and he embraced animal observation as a dedicated course of study. Tinbergen was inspired by the
work of Karl von Frisch to study the homing abilities of the digger
wasp, on which he wrote his thesis. Having "scraped through my finals
without much honour," Tinbergen, along with his young family, next

---

36 K. Lorenz, Banquet Speech, December 10, 1973, NobelPrize.org, nobelprize.org/
prizes/medicine/1973/lorenz/speech/.

37 W. Sullivan, "Questions Raised on Lorenz's Prize," *New York Times*, December 15,
1973.

38 As quoted by W. Cloud, "Winners and Sinners," in *The Sciences*, December 1973,
published by the New York Academy of Sciences, doi.org/10.1002/j.2326-1951.1973.
tb01360.x.

39 "Nobel Prize Laureate Posthumously Stripped of Academic Honors for Embracing
Nazism," Associated Press, December 17, 2015.

spent more than a year as European observers in a primitive Eskimo village in Greenland.[40]

Tinbergen and Konrad Lorenz's joint publications described instinctual behaviors as fixed-action patterns consisting of a repeated and distinct set of movements or behaviors triggered by events in the animal's environment. They asserted that these characteristics were shared by all members of a species, including humans. Recalling Lorenz's explanation of cat behavior, Tinbergen made the same point with dogs: "As every dog-owner knows, a dog can by no means be prevented from making hunting excursions by supplying it with ample food. Other instances of a similar kind are familiar to us by introspection. Sports, science and so many other activities certainly have connections with internal factors of this kind."[41] And so "by introspection," humans are to conclude that endeavors such as sports and science are likewise instinctual, caused entirely by some series of brain mechanisms.

Tinbergen's subsequent detailed expositions of animal behaviors corroborated most of Lorenz's theories with only occasional minor differences. Although he did agree with Lorenz that human cultural influences were most of what kept instinctual behaviors in check, Tinbergen voiced more awareness of the dangers inherent in applying animal data to human behavior.

At the outbreak of WWII, Tinbergen was on the faculty at Leiden University. Early in the war, Holland had the idea that the country could remain neutral, and by doing nothing to incite Hitler, had hoped to remain unmolested. To this end, the Dutch repeatedly declined requests from Britain and France to join the Allies, while at the same time putting their border troops on alert for a possible German invasion. The Nazis rolled into the Low Countries in May 1940 and

---

40 Nikolaas Tinbergen, Biographical, NobelPrize.org, circa 1973, nobelprize.org/prizes/medicine/1973/tinbergen/biographical/.

41 N. Tinbergen, *An Objectivistic Study of the Innate Behaviour of Animals* (Leiden: Brill, 1942).

overcame the weak Dutch forces in a mere four days. While Nazi brutality toward Jews started immediately, there was a stepwise approach to oppressing the non-Jewish population as well. In 1942, Tinbergen joined about 80 percent of faculty members at the University of Leiden in signing a letter in protest of the Nazi presence. They were immediately arrested, and Tinbergen spent two years in a prison camp, by all accounts in relative comfort. Tinbergen moved his family to England after the war for a position at Oxford University, where he stayed for the rest of his career.

An example of Tinbergen's animal observations includes a description of the environmental trigger of a fish called the three-spined stickleback. During mating season, the males of the species turn bright red and blue, and rival males attack each other. Through a series of experiments, Tinbergen discovered that the trigger for aggression was seeing the red spot on a rival's belly. When Tinbergen taunted males with dummy objects—not even fish shaped but with a red spot—they launched into aggressive behaviors. Thus Tinbergen identified it was solely the color red that provoked the response.[42]

Tinbergen's Nobel lecture did not concern any of his work on animal behavior and instead focused on two separate human topics that occupied his later career: autism and the Alexander technique. Surprisingly, Tinbergen rejected the genetic and brain-based theories of autism, instead considering it was entirely due to something that went terribly wrong in early childhood social interactions.[43] This could be a frightening accident or other sudden shock or (his favorite theory) a mother who was not sufficiently demonstrative in her protective behavior toward the infant. Tinbergen described success with a gentle gradient technique that, first of all, did not insist on eye contact with

---

42 N. Tinbergen, "The Curious Behavior of the Stickleback," *Scientific American* 187, no. 6 (1952): 22–27; N. Tinbergen, *The Study of Instinct* (Oxford: Oxford University Press, 1951).

43 E. A. Tinbergen and N. Tinbergen, *Early Childhood Autism: An Ethological Approach* (Berlin: Parey, 1972).

the child, and secondly, allowed the child to make physical contact. This was followed by very gradual increases in communication of the very mildest nature, including repetition of sounds elicited by the child and quiet acknowledgments of anything the child offered to the adult. Tinbergen also endorsed working with the mother to convince her to engage the child with protective behavior.

Tinbergen's other lecture topic, the Alexander technique, was named after its creator, Frederick Matthias Alexander, a nineteenth-century Shakespearean orator from Tasmania who developed the method in order to treat his own disability. Alexander had nearly lost his voice, and by observing himself in a mirror, he discovered that he had gotten into the habit of adopting abnormal and unnecessary postures in an effort to force vocalizations. His technique consisted of reeducating himself on the relaxed posture and correct body movements that would result in an effortless, proper way to make sounds. This developed into additional techniques to improve spatial awareness and body movement in general, with alleged benefits on mental and emotional states.[44] Tinbergen himself experienced benefits of the Alexander technique, and applied his ethological approach to analyzing why it might work. He described that what have been called psychosomatic ills might instead be "somato-psychic." For example, if one is in the habit of slouched sitting, one can begin to feel insecure— for those in fear adapted a cowed posture.[45]

Nikolaas Tinbergen died in 1988 at the age of eighty-one, following a stroke.

---

44 F. M. Alexander, *The Use of Self* (London: Chaterston, 1932).

45 N. Tinbergen, "Ethology and Stress Diseases," December 12, 1973, NobelPrize.org nobelprize.org/prizes/medicine/1973/tinbergen/lecture/.

24

Inside the Cell

Albert Claude, Christian de Duve, and George Palade shared the 1974 Nobel Prize for their discoveries of the inner structures and organization of the cell.

Albert Claude was born in 1899 in Longlier, Belgium, a sparsely populated area of extensive forests, rough terrain, and rolling hills. Claude and his family spent their summer evenings outdoors with an uninterrupted view of the Milky Way. As a youth, he was a bell ringer at the centuries-old local cathedral, until the dilapidated structure came crashing down one night in a storm.

Albert Claude's early years were spent at the side of his mother until she died of cancer when he was seven. He attended a one-room school with children of all ages, which he considered to have been an excellent educational setting. This was followed by the family's move to Athus, where the local school lessons were all in German. His education was interrupted when young Albert was recalled to Longlier to care for an uncle who had suffered a stroke.

In the early twentieth century, the Ardennes region was transforming from a logging and farming community to a center for metalworking, and as a teenager, Albert was apprenticed as a draftsman at a steel mill. At the outbreak of WWI, one of Germany's early moves was to occupy Belgium, prompting Claude to join the British intelligence service. Belgian operatives monitored troop movements, especially by train transports, and airfield activity; identified the locations of ammunition depots; and interrogated German deserters. Many of their communications were made via carrier pigeon. They facilitated distribution of a clandestine newspaper, operated a covert mail system, and aided people escaping to the Netherlands and England. The Germans were ruthless when they caught intelligence agents, and many were executed. After the war Claude was assigned veteran status and decorated with the Interallied Medal.

Claude was inclined to train in metallurgy, but he changed direction when a government initiative allowed for those with war-interrupted educations to enroll in university despite not possessing high school diplomas. He took the opportunity to study medicine. "The medical school was the only place where one could hope to find the means to study life, its nature, its origins, and its ills."[1] Even as a premedical student, Claude worked in the lab with keen interest in discerning the secrets of the cell. He earned a medical degree from Université de Liège in Belgium in 1928.

In 1929, Claude determined that he would work at the Rockefeller Institute for Medical Research in New York City. He boldly crafted a letter in imperfect English to its director, Simon Flexner, proposing to characterize the sarcoma virus. Although this virus had first been

---

1   A. Claude, Biographical, circa December 1974, NobelPrize.org, nobelprize.org/prizes/medicine/1974/claude/biographical/.

identified as a cause of chicken tumors in 1911 by Peyton Rous, its existence and character as a virus was still seriously doubted. Flexner accepted him, and Claude moved to the US in 1929, becoming a citizen in 1941.

In his efforts to isolate and purify the sarcoma virus in tumor cells, Claude realized that before characterizing cancerous cells, he really needed to know what normal cells looked like and how they functioned. Claude utilized the recently developed technique of crushing cells with a mortar and pestle then spinning them in a centrifuge. This would cause the various cell constituents to settle into clean layers, with heaviest components on the bottom, lighter ones higher up, and the free fluid of the cell in the top layer. Claude worked to perfect these techniques and succeeded in not only purifying the Rous sarcoma virus but also identifying that healthy cells contained discrete structures that could be separated by this method. He conducted many series of biochemical tests on the layers to discover their functions in the cell.

Large structures within cells had been visualized by use of the light microscope in 1894, but investigations during the subsequent fifty years had not determined their design or function. Albert Claude provided the breakthrough experiments that identified that these intracellular entities, now called mitochondria, were the real power plants of the cell. Years later, Claude would wonder at his luck that the harshness of his early cell-preparation methods didn't destroy the cell components completely.

Claude and his coworkers published their findings in 1943, which prompted an invitation from the New York company Interchemical Laboratory to use their electron microscope, one of the few in the country at the time.[2] Claude's team published the first image of an

---

2   A. Claude, "The Constitution of Protoplasm," *Science* 97 (1943): 451–56.

intact cell and mitochondria in 1945.[3] It is for these discoveries that Albert Claude was recognized with the Nobel Prize in 1974.

Claude returned to Belgium in 1950 and built a cancer research center at the University of Belgium. After two decades of service at the university he retired from cancer research and headed the Institut Jules Bordet in Brussels, named in honor of the Nobel Prize winner from 1919.[4]

Cell biology had established that tiny cells form the basic constituents of all living organisms. Along with many others, Claude speculated about how cells carried genetic information: "In addition, we also know that the cell has a memory of its past, certainly in the case of the egg cell, and foresight of the future, together with precise and detailed patterns for differentiations and growth, a knowledge which is materialized in the process of reproduction and the development of all beings from bacteria to plants, beasts, or men."[5]

Albert Claude took the opportunity of his Nobel lecture to comment on the reason he had entered the medical research field: to find the source of life. He acknowledged the role of chance and probability yet couldn't reconcile a theory that invoked pure chance with the spectacular intricacies, supreme coordination, and remarkable efficiencies of cellular life that he had helped to discover. He asked, "In the name of the experimental method and out of our poor knowledge, are we really entitled to claim that everything happens by chance, to the exclusion of all other possibilities?"[6]

Albert Claude was a lively optimist, and he closed his Nobel lecture with this uplifting promotion of the future: "Let us trust that mankind,

---

3   A. Claude and E. F. Fullam, "An Electron Microscope Study of Isolated Mitochondria: Methods and Preliminary Results, *Journal of Experimental Medicine* 81 (1945): 51–62.

4   For more information on Jules Bordet, see *Boneheads & Brainiacs*.

5   A. Claude, "The Coming Age of the Cell," December 12, 1974, NobelPrize.org nobelprize.org/prizes/medicine/1974/claude/lecture/.

6   A. Claude, "The Coming Age of the Cell," December 12, 1974, NobelPrize.org nobelprize.org/prizes/medicine/1974/claude/lecture/.

as it has occurred in the greatest periods of its past, will find for itself a new code of ethics, common to all, made of tolerance, of courage, and of faith in the Spirit of men."[7]

Albert Claude died 1983 at the age of eighty-four.

———————

Christian de Duve was born in in 1917 in Surrey, outside of London, where his parents had taken refuge from the German occupation of Belgium. The family returned home after the war, and de Duve grew up in a multicultural atmosphere, yearly visiting friends the family had made in England and regularly vacationing at homes of relatives in Germany, all the while being educated in Flemish and French at a Jesuit school in Antwerp. He could read books before the age of five and became multilingual early on. De Duve entered the Catholic University of Louvain at seventeen and there developed more interest in the humanities than the sciences, yet he held a romantic image of "the man in white healing the sick" that attracted him to study to become a physician.[8] In medical school he spent his free time in the lab to catch up on his science studies, and that was where he discovered the power of experimentation to get answers, rather than the routes of philosophy and theology in which he had been educated.

In the labs at Louvain, de Duve experimented on dogs to determine the role of the liver in insulin metabolism. This was a remarkable course of investigation at the time, since the world's most prominent researchers had firmly established that insulin's main effect was on the muscles, causing them to take up sugar for energy. The Louvain group showed the liver also had important roles in glucose metabolism, although their findings remained disputed for many years. It is

———————

7   A. Claude, "The Coming Age of the Cell," December 12, 1974, NobelPrize.org nobelprize.org/prizes/medicine/1974/claude/lecture/.

8   Interview with C. de Duve, "Christian de Duve–Choosing a Career: The Decision to Become a Physician (7/106)," video, July 11, 2017, Web of Stories, youtube.com/watch?v=1dznz93UmAk.

now known that the liver has the primary role in how the body deals with sugar.

De Duve witnessed the rise of the Nazi government, and in 1939, he saw the Nazi leader Hermann Goering swimming at a vacation lake shared with the German side of de Duve's extended family. At the outbreak of World War II in September 1939, German bombers struck the city of Louvain, targeting the new library in order to deliver a harsh message. The library had been burned to the ground in WWI by Germany and had since been rebuilt, representing a symbol of civilization against barbarism. De Duve promptly enlisted in the Belgian army and was sent to France, where his unit was apprehended by the Germans within a few weeks.

While on a forced march under loose guard, de Duve and his friends opted to stop at a café. He did not appreciate the beatings received when the Germans retrieved them, and at his next opportunity on the march de Duve jumped onto the running board of a passing car and held on for dear life. Thus he escaped captivity and found his way back to Belgium.

In occupied Belgium, de Duve stayed under the radar of the occupying Germans for the duration of the war, managing to complete his medical degree at Louvain in 1941. By that time, de Duve had decided against a career in patient care and opted to become a research scientist. His efforts to gain more experience with chemistry were frustrated by the fact that lab work had virtually stopped during the war, owing to difficulties in getting equipment and funding. De Duve managed to find a job as a hospital assistant at the local cancer institute, where the basement was full of laboratory rodents. He carried out the director's crude research, which consisted of giving cancer-causing substances to separate groups of rats, then trying out various interventions to see if cancer could be prevented, such as putting them on vitamins or special diets or feeding them metals. De Duve's job was to follow the incidence of cancers that developed.

De Duve managed to study some chemistry thanks to funding from a start-up drug company. His thesis work consisted of developing a quick and inexpensive way to purify penicillin. This product led to the creation of the RIT pharmaceutical company, which grew into a huge corporation and was eventually bought out by the American company Smith, Kline & French. After attaining the equivalent of a master's of science in chemistry, de Duve still felt he lacked adequate biochemistry skills. He wrote to Hugo Theorell (who would win a Nobel Prize in 1955) at Karolinska University in Stockholm to ask for a job. Although de Duve admitted his lack of experience, Theorell took him in as a lab assistant and put him to work isolating human hemoglobin in the form of crystals. In the course of this work, de Duve learned valuable experimental techniques. De Duve fondly recalled the antics of Theorell, who was partially paralyzed from childhood polio and walked on crutches: "He could walk on his hands better than his feet and sometimes he would walk around the lab like that singing some opera tune!"[9]

Because Theorell was a member of the Nobel Committee, the students in the lab, including de Duve and his wife, were invited guests at the Nobel festivities in 1946. At the end of ceremonies at two in the morning, they emerged into the bitter cold to catch a bus back to their apartment. When his wife said she wished she had a fur coat de Duve replied, "You will get a fur coat when I get a Nobel Prize."[10] She got her fur coat in 1974!

De Duve had a knack for attaching himself to brilliant researchers, and once again he wrote to future laureates, this time Gerty and Carl Cori in St. Louis, Missouri. He asked for a six-month job in order to gain experience in the area of insulin research. Shortly after de Duve arrived in St. Louis, the Coris were awarded the Nobel Prize. In St. Louis, de Duve was assigned to work with yet another future laureate,

9   Interview with C. de Duve, "Christian de Duve–Sweden, and Memories of Hugo Theorell (18/106)," video, July 11, 2017, Web of Stories, youtube.com/watch?v=JGlhn6cOl20.

10 Inerview with C. de Duve, "Christian de Duve–'You will get a fur coat when I get a Nobel Prize' (23/106)," July 11, 2017, Web of Stories, youtube.com/watch?v=tcs50aYVm_U.

Earl Sutherland, with whom he made an important discovery on the hormones controlling sugar metabolism.

De Duve returned to Louvain to continue his original insulin research, but he became sidetracked when using a new technique to sort chemicals, namely the centrifugation methods that had been so successful for Albert Claude. When Louvain centrifuged cells, he found that one of the layers was composed of small round membrane-bound packets that were highly acidic inside. When he investigated, he found these acid bags to have the function of internal digestion, acting like a stomach of the individual cell. De Duve named these lysosomes, from *lyse*, "to break up," and *soma*, "body." They float freely within the cells outside the nucleus, and ingest and dissolve unwanted parts of the cell, cell debris, or foreign substances that have entered the cell. The acidic interior and digestive enzymes of the lysosomes break down large structures and molecules into simple components and then return the products to the cell for further use or disposal. So lysosomes act as a cellular stomach plus a recycling system in one. De Duve won the Nobel Prize for this work by his team at Louvain.

Research on lysosomes revealed that every nucleated cell has them, and they have specialized functions depending on the tissue. For example, the lysosomes in kidney cells digest filtered proteins, while in the thyroid gland, lysosomes digest thyroxine hormone. De Duve discovered that a more general function of lysosomes is to release their digestive juices directly into the cell to cause the demise of the cell, in other words, cellular suicide or, properly termed, *autophagy* ("self-eating"). Autophagy will only occur if there is already some cellular damage.

The discovery of lysosomes revealed the cause of some rare diseases that are collectively called lysosome-storage diseases. Individuals may inherit genetic variants that render their cells unable to produce lysosomal enzymes, preventing their lysosomes from being effective processors of cellular garbage. The cellular debris is allowed to accumulate and causes cell death. Most affected persons die in childhood.

Lysosomes and COVID

The discoveries of the 1974 Nobel Prize winners are now the stuff of routine memorization for medical students. After learning the normal function of lysosomes and their dysfunction in rare genetic diseases, practicing physicians don't devote much everyday thinking to them. That changed in 2020, when new studies on drugs to treat COVID sent many infectious-disease doctors scurrying back to their basic cellular-biology textbooks.

A report from the Karolinska Institute in Stockholm showed that drugs that selectively accumulate in lysosomes can be effective against the SARS-CoV-2 virus that causes COVID. The researchers systematically evaluated 530 existing drugs to see if they accumulate inside lysosomes. Once inside the lysosome, the drugs, called lysosomotropic agents, become acidic and cannot easily exit the lysosome. They cause the lysosomes to enlarge. Lysosomotropic drugs serve to downregulate autophagy. This alone can help a cell survive damage from a virus. These drugs also decrease the permeability of the cell membrane to invading viruses. Along with other biochemical actions, these drugs impair the success of the virus in utilizing the cell's machinery in making more copies of virus.

The researchers identified thirty-six lysosomotropic drugs with possible antiviral effects, including fourteen that already had proven antiviral effects. These include azithromycin (traditionally classified as an antibiotic); chloroquine, hydroxychloroquine, and quinine (traditionally classified as antimalarials); formoterol (traditionally used to dilate the airways); and meclizine (classified as an antihistamine), among other compounds.[11]

---

11 U. Norinder , A. Tuck, K. Norgren, and V. Munic Kos, "Existing Highly Accumulating Lysosomotropic Drugs with Potential for Repurposing to Target COVID-19," *Biomedicine & Pharmacotherapy*, 130 (October 2020): 110582.

On hearing that he won the Nobel Prize, de Duve had mixed emotions. He was extremely happy and gratified, but also found it disturbing because it disappointed those who were not selected to share the prize, even though they had done as much work. De Duve admitted these included many of his coworkers and friends. "It's very much like winning a lottery. The only thing is the tickets are expensive."[12] De Duve said that all researchers dream of winning the Nobel Prize but that cannot be the sole impetus for their work. "The Nobel Prize is not like winning a race. You just try to do the best work you can and if you're lucky and if you're recognized by your peers—which doesn't always happen—you may get a prize like this."[13] De Duve subsequently experienced that the prestige of the prize created an uncomfortable artificial separation from nonwinners, and he tried to overcome this in his professional associations with colleagues on both sides of the Atlantic.

The first thing he had to do with his winnings was to buy his wife a fur coat, as promised in 1946. Then he bought himself a new piano and built a swimming pool—an investment he and his wife were still using at an advanced age.

De Duve led a transcontinental life, holding a professorship at Louvain while also being a professor at Rockefeller University in New York City. He continued in research with many more discoveries, and one of his greatest contributions was to write an illustrated two-volume cell-biology book for high school students.[14] Although some of the concepts have since been refined, it is still an excellent resource for the beginner.

In retirement de Duve turned his attention to the theories of the origin of life on Earth. His various books trace what is known and not known about the stages of life development, most of it of course

---

12 Interview with C. de Duve, "Christian de Duve–Learning I'd won the Nobel Prize (61/106)," video, July 11, 2017, Web of Stories, youtube.com/watch?v=Yu8_xnnopB0.

13 As quoted in N. Hicks, "3 Nobel Laureates in Medicine," *New York Times*, October 11, 1974.

14 C. de Duve, *A Guided Tour of the Living Cell* (New York: W. H. Freeman & Co., 1984).

being *not* known, because primitive life matter is not preserved in rocks or the atmosphere. De Duve conjectured about a large number of discontinuous, unique events that might have occurred, each of which, he supposed, was pivotal in facilitating the cascade of events that culminated in life. He called these "singularities," because they only ever happened once, and as such they cannot be experimentally verified.[15] There were a lot of these singularities, according to de Duve. Creationists easily identify these remarkable events as miracles. Materialists like de Duve attribute the singularities to key chemistry reactions taking place in exceedingly lucky instances in just the right environment, while admitting that when a soup of chemicals, heat, and acid are created in the presumed "right" laboratory environment, they fail to spark life.

In summary, de Duve said, "The answer of modern molecular biology to this much-debated question is categorical: chance, and chance alone, did it all, from primeval soup to man, with only natural selection to sift its effects."[16] Somewhat later in life, de Duve sounded less definite, admitting, "The problem of the origin of life is easy to state, if not solved."[17]

For decades de Duve held a leadership position at the Catholic University of Louvain while refraining from expressing his views as a nonreligious person. Finally he broke his silence in 2002 by writing the book *Life Evolving*, which was directed to his friends and Louvain colleagues, in which he admitted that he thought religion was nonsense.[18] De Duve was one of forty-one Nobel laureates who endorsed a campaign to repeal a Louisiana law that permitted public schools to

15 C. de Duve, *Singularities: Landmarks on the Pathways of Life* (Cambridge: Cambridge University Press 2005).

16 C. de Duve, *A Guided Tour of the Living Cell*, vol. 2 (New York: Scientific American Library, 1984), 357.

17 Interview with C. de Duve, "Christian de Duve–LUCA and RNAs (102/106)," video, July 11, 2017, Web of Stories, youtu.be/DeS6g9Y05D4.

18 C. de Duve, *Life Evolving: Molecules, Mind, and Meaning* (New York: Oxford University Press, 2002).

allow teaching materials on the various theories of the origin of life that are counter to established scientific theories. Of course theories of the origin of life are the very content of several of de Duve's own books. The campaign failed.

Teaching the Origins of Life

The Louisiana Science Education Act was proposed by twenty-six state senators and sixty-six state representatives and signed into law in 2008. It allows public school teachers to use supplemental materials in the science classroom that are critical of scientific theories and discuss ethics on such topics as evolution, global warming, and human cloning. It has been characterized by some as violating the federal statute prohibiting the teaching of religion in public schools. However, the act specifically states that school officials are to "foster an environment within the public elementary and secondary schools that promotes critical thinking skills, logical analysis, and open and objective discussion of scientific theories being studied including, but not limited to, evolution, the origins of life, global warming, and human cloning."[19]

In 2010, a high school student launched a campaign to repeal the law. The repeal effort was sponsored by a state senator and endorsed by forty-one Nobel laureates, including Christian de Duve. Repeated legislative proposals to repeal the act failed to advance past the Senate Education Committee annually from 2010 to 2015.

As he advanced in age, de Duve developed a poor impression of humankind and despaired over the gloomy future of Earth. He felt that while lucky "singularities" enhanced short-term survival, they also had a downside: "The negative counterpart of those 'good' traits has been defensiveness, distrust, competitiveness, and hostility toward

---

19 Act no. 473, 2008 Regular Session of the Louisiana Legislature, legis.la.gov/Legis/ViewDocument.aspx?d=503483&n=SB733%20Act%20473.

the members of other groups, the seeds of the conflicts and wars that landmark the entire history of humanity up to our day." He further lamented, "In spite of the advances of medicine, deathly epidemics are more menacing than ever before. Conflicts, exacerbated by economic disparities, nationalisms, and fundamentalisms, are raging in various parts of the world. The specter of a nuclear holocaust has become thinkable."[20] He urged population control and suggested that genetic engineering could someday modulate some of mankind's more "destructive impulses."

With this cynical view of his fellows, a pessimistic outlook for the future, and no religion to buffer his dark thoughts, it is no wonder that at the age of ninety-five Christian de Duve elected for euthanasia by lethal injection, citing failing health. "It would be an exaggeration to say I'm not afraid of death, but I'm not afraid of what comes after because I'm not a believer. When I disappear I will disappear, there'll be nothing left," he told the Belgian daily *Le Soir* a month before his suicide.[21] Christian de Duve died in 2013.

The third winner of the Nobel Prize in 1974 was George Emil Palade, who was the last of the trio to embark on intracellular research. In 1946, Palade was a recently arrived Romanian war refugee working as a research physician in the labs at New York University. He attended a symposium held at the university where he heard a presentation by Dr. Albert Claude on the use of the electron microscope to study cells. Palade immediately approached the speaker to arrange to meet in his laboratory, and thus Palade became Claude's student and colleague.

20 C. de Duve, *Genetics of Original Sin: The Impact of Natural Selection on the Future of Humanity* (Paris: Odile Jacob, 2009).

21 D. Gellene, "Christian de Duve, 95, Dies; Nobel-Winning Biochemist," *New York Times*, May 6, 2013.

Palade was born in 1912 in Iasi, Romania.[22] After finishing medical training, he was drafted into the Romanian army to serve as a physician. Palade did not speak publicly or write about his war experience, but it must have been difficult. At the onset of World War II in September 1939, Romania declared its neutrality. In September 1940, the country was taken over by a military dictatorship which embraced an alliance with Germany. Romania declared war on its giant neighbor and joined the German invasion of the Soviet Union. The Soviets responded by bombing Romania.

Meanwhile the police and Romanian and German army personnel, assisted by Romanian civilians, carried out a brutal holocaust second only to Germany's in magnitude.[23] On June 29, 1941, there was a massacre of an estimated thirteen thousand Jews in Palade's hometown of Iasi, followed by train cars being loaded with thousands more people being transported to nearby concentration camps. In reality the trains were made to be death containers. They were sealed up and progressed slowly over one week in the hot summer, going back and forth on the same track before arriving the few kilometers to their destination. Most perished on the death trains, which made one stop to be "checked" by doctors and the Red Cross, who could only offer water before the cars were resealed and moved slowly on. In 1943, the Soviet army advanced on Romania as the war turned unfavorable for Hitler, and Romania was bombed by the Allies. The Soviets occupied Romania in August 1994, at the same time that another coup deposed the military leader, and the country joined the Allies.

In 1944, Palade was desperate to escape communist rule. Having sent his wife and daughter ahead to Turkey, Palade managed to secure a fake visa and passport on the black market and sneak out of the

---

22 Iasi is also known as "Jassy."

23 "Executive Summary: Historical Findings and Recommendations," in *Final Report of the International Commission on the Holocaust in Romania,"* International Commission on the Holocaust in Romania, November 11, 2004, jewishvirtuallibrary. org/jsource/Holocaust/Romania/execsum.pdf.

country in the middle of the night. He met his family in Turkey and brought them to Casablanca, Morocco. It wasn't until the end of 1945 that they obtained passage to America. Palade's only resources were several letters of recommendation addressed to various US labs written by his professor in Romania, and through these, he secured a research job at New York University.

Albert Claude was impressed by Palade's interest and offered him a job in his lab at the Rockefeller Institute for Medical Research. Palade worked with Claude's team to isolate mitochondria from liver cells, to stain them in order to see their internal structure, and to conduct various tests to determine their function.[24] Palade developed a method to preserve the mitochondria in a solution that did not disrupt its native architecture.[25] The improved techniques brought tremendous clarity to the structure of intracellular components and established the highest cell-preparation standards in the world of cell biology.

Palade was most interested in the bridge between structure and function. He continued intracellular research, making contributions in identifying several intercellular structures, the most famous of which was named after him for a short time—"Palade's particles" soon came to be known as ribosomes.[26] Ribosomes are granule-looking structures found in all living cells. They are actually complex molecular machines that make all of the body's proteins. DNA contains the code for making proteins, which is transcribed into messenger RNA, and then translated by the ribosomes. Transfer RNA brings the requisite amino acid building blocks to the ribosome. The ribosome then has the blueprints and the bricks to construct the proteins.

24 G. E. Palade, "The Fine Structure of Mitochondria, *Anatomical Record* 114 (1952): 427–51.

25 G. Palade, "A Study of Fixation for Electron Microscopy," *Journal of Experimental Medicine* 95, no. 3 (1952): 285–98.

26 M. Farquhar, "A Man for All Seasons: Reflections on the Life and Legacy of George Palade," *Annual Review of Cell and Developmental Biology* 28 (November 2012): 1–28.

In 1970, Palade married his longtime collaborator Dr. Marilyn Farquhar, also a cellular biologist of considerable note. They were recruited to Yale University Medical School in 1973 to develop the section of cell biology there, which was dear to Palade's goals of seeing basic research translated into actual patient care. In 1990, at the age of seventy-seven, Palade became the dean of science in the School of Medicine at UC San Diego, where he continued to research in addition to his administrative duties. He served in a reduced capacity even after formal retirement in 2000. George Palade enjoyed poetry, impression-istic art, and attending live music performances. He regularly took part in vigorous mountain hiking, and he loved socializing with family and friends.

George Palade died in 2008 at the age of ninety-five.

25

Infected by Viruses

The 1975 Nobel Prize in Medicine was shared by Renato Dulbecco, Howard Temin, and David Baltimore for their discoveries on how tumor viruses interact with genes in the cell.

Renato—literally, "the reborn"—started his life in 1913 and was named by a mother who had previously lost a baby boy to meningitis. Renato Dulbecco's mother was a strong influence, teaching him to read and do math when he was only three years old. This helped him to speed through school and graduate at the age of sixteen. Although he became adept at piano and was fascinated by opera, his mother urged Renato to study medicine. He graduated from medical school at twenty-two.

Dulbecco worked in the anatomy labs of Giuseppe Levi, a teacher of three eventual Nobel laureates from Italy, including Salvador Luria and Rita Levi-Montalcini.[1] Dulbecco learned how to grow cultures of cells in the lab, but he credits Levi with far more than imparting exacting technical standards, describing his mentor as encouraging free thinking in postulating new directions for research coupled with an attitude wide open to scrutiny and criticism.

---

1   Rita Levi-Montalcini won the prize in 1986 for the codiscovery of nerve growth factor.

Dulbecco's studies were interrupted by mandatory army service for two years, but he had only been back in the lab briefly when he was reconscripted into the Italian brigade supporting Germany's advance into Russia. On their way to the Russian front, his troop-transport train paused at a rail yard in Nazi-occupied Poland, where Dulbecco observed Jews, identified by yellow patches on their clothing, working to repair the tracks under German guards. He inquired what was going on and was gleefully informed by a German that the Jews would be murdered after they completed their labors. Dulbecco was stunned and told his fellow Italians who, like him, had no prior inkling of the nature of Nazi atrocities. He made a decision at that moment and determined that if he survived his military service, he would not have anything to do with the army in the future. Dulbecco served as a medic one hundred meters behind the front lines in Russia, where his regiment lost 80 percent of their men, among them many of his childhood friends. Dulbecco was injured in a bomb blast and transferred to a hospital for nine months; he was ultimately determined to be unfit for further duty and put on a train to home.

Dulbecco moved his young family to a small town and worked as a doctor, keeping his head down while joining various resistance groups, first the Christian Worker's Movement (a splinter group of the Communists) and then the Turin Liberation Committee. At that time, the country was under the dictatorship of Mussolini, but it was fully occupied and virtually run by the Nazis. Throughout the occupation, Dulbecco provided medical assistance to resistance workers and taught first aid to volunteers. In 1943 the war turned against Hitler. Dulbecco recalled standing by the hospital in Turin, fully prepared for casualties in anticipation of violence as the Germans withdrew, but all that he heard for hours and hours was the rumble of tanks and trucks exiting in a hurry. Mussolini was ousted as the Italians came under the influence of the Allies.

After the war, Dulbecco worked again in the Levi lab, where he found out that his colleague Levi-Montalcini was headed to St. Louis,

Missouri, to work with Gerty and Carl Cori.[2] Dulbecco was visited by Salvador Luria, who invited him to his lab at Indiana University.

Dulbecco took a ship to New York, where he was met on the dock by a relative of a friend, who informed him that all of the hotels were full and dropped him off at a Turkish bathhouse. There one could sleep the night in a private cubicle provided to bathers for relaxation after immersion. (This was at a time before bathhouses earned an unsavory reputation.) The next day Dulbecco boarded a train for Bloomington, Indiana, and he was soon joined by his family from Italy.

At Indiana, Salvador Luria had recently discovered that viruses that infect bacteria (bacteriophages) could be inactivated by ultraviolet light. This led to Renato Dulbecco's first significant discovery, as he conducted various tests to see if anything would revive the inactivated viruses. He quickly discovered that exposure to visible light caused a prompt reactivation of the bacteriophages.[3] This drew the attention of future Nobel laureate Max Delbrück at Caltech, who invited Dulbecco to work in his lab. Shortly after his arrival, Caltech was endowed with a generous grant to study animal viruses, and Dulbecco volunteered for the project, for which he was paired with Delbrück's lab researcher Marguerite Vogt. The first major contribution to this field by Vogt and Dulbecco was to devise animal-cell-growth techniques that could allow exact counting of infections from viruses.[4] The two proceeded to perfect these techniques for many different kinds of animal viruses.[5] A

---

2 For more on Gerty and Carl Cori, see *Boneheads & Brainiacs*, 187–91.

3 R. Dulbecco, "Reactivation of Ultra-Violet-Inactivated Bacteriophage by Visible Light," *Nature* 163 (1949): 949.

4 R. Dulbecco, "Production of Plaques in Monolayer Tissue Cultures by Single Particles of an Animal Virus, *Proceedings of the National Academy of Sciences of the United States of America* 38 (1952): 747–52.

5 R. Dulbecco and M. Vogt, "Some Problems of Animal Virology as Studied by the Plaque Technique," *Cold Spring Harbor Symposia on Quantum Biology* 18 (1953): 273–79; R. Dulbecco and M. Vogt, "Plaque Formation and Isolation of Pure Lines with Poliomyelitis Viruses," *Journal of Experimental Medicine* 99 (1954): 167–82; R. Dulbecco and M. Vogt, "One-Step Growth Curve of Western Equine Encephalomyelitis Virus on Chicken Embryo Cells Grown In Vitro and Analysis of Virus Yields from Single Cells," *Journal of Experimental Medicine* 99 (1954): 183–99.

few years later, the student Howard Temin joined Delbrück's lab, where Temin perfected the technique for counting Rous sarcoma viruses in cell culture.[6] These techniques led directly to the development of the Salk polio vaccine and later to the use of lab cell cultures to test various substances for their cancer-causing potential.

Vogt and Dulbecco observed that sometimes the viral DNA simply replicates until it bursts the host cell, while at other times it transforms the host into a cancerous cell.[7] In 1962, Dulbecco and Vogt moved to the Salk Institute, where they continued their collaboration and discovered that the viral DNA that they were working with was in a circular shape. This was significant because circular viral DNA could be copied into cellular DNA without losing any code material.[8] About five years after this observation, Dulbecco and others in the lab demonstrated that the viral DNA does indeed become incorporated into the host cell's DNA.[9] Every time the cell divides to make more cells it imparts the viral DNA to its progeny, and all of the cell generations after the infection carry the viral DNA and the programming for cancer.

There are only a handful of known viruses that have the potential to cause cancer in humans. The value of Dulbecco's work is that the cancer-causing viruses provide an ideal model in which to study the genes that control transformation into cancer, cancer-cell growth, and cancer-cell death. In his attempts to discover what determines whether a cell will simply be killed by a virus or will turn cancerous because of incorporated viral DNA, Dulbecco realized the research was stalled

---

6   H. Temin and H. Rubin, "Characteristics of an Assay for Rous Sarcoma Virus and Rous Sarcoma Cells in Tissue Culture," *Virology* 6, no. 3 (1958): 669–88.

7   M. Vogt and R. Dulbecco, "Virus-Cell Interaction with a Tumor-Producing Virus," *Proceedings of the National Academy of Sciences of the United States of America* 46, no. 3 (1960): 365–70, doi.org/10.1073/pnas.46.3.365.

8   R. Dulbecco and M. Vogt, "Evidence for a Ring Structure of Polyoma Virus DNA," *Proceedings of the National Academy of Sciences of the United States of America* 50, no. 2 (1963): 236–43.

9   R. Dulbecco et al., "The Integrated State of Viral DNA in SV40-Transformed Cells," *Proceedings of the National Academy of Sciences of the United States of America* 60: 1288–95.

due to only knowing a few parts of the entire genetic codes, not only of viruses, but also of the bacteria under study and of the animals and humans they infect. He became a champion of the Human Genome Project in the US and led a parallel effort in Italy. It has since been discovered that there are hundreds of host-cell factors that determine how active a viral portion of DNA will be inside the host cell, which has led to new drug treatments for cancer.

Meanwhile, Dulbecco's attention turned toward the study of breast cancer, work that he started in 1972 when he took a break from the Salk Institute to work at the Imperial Cancer Research Fund Laboratories in London. Dulbecco was living in the posh suburb of Chislehurst in Kent in England in 1975 with his new wife and young daughter when he was notified of his Nobel Prize. He told a humorous story of the local reaction to the announcement: "How is it possible for someone from here to win the Nobel Prize! Because they thought they were all rich, and rich people didn't work."[10]

Marguerite Vogt

Renato Dulbecco's autobiographical note on the Nobel website mentions Marguerite Vogt, and his Nobel lecture cites seven papers written by or with Vogt. She was a German-born researcher who published her first scientific paper—on genetic mutations in the fruit fly—at age fourteen. She came to the US to work at Caltech in 1950, bringing only her grand piano. Vogt disdained to speak German and had no desire to visit Germany, gladly distancing herself from the Nazi era in which she grew up. Vogt was not only at the forefront of the early discoveries with Dulbecco at Caltech, she continued to make important contributions to the understanding of cancer and viruses in her impressive eight-decade career. Her

---

10 Interview with R. Dulbecco, "Renato Dulbecco – 'Nobel coming' (32/61)," November 1, 2017, video, Web of Stories, youtube.com/watch?v=Jgrp4z1X5Vo&list=PLVVOr6C mEsFwzIooOixRbz6qqo-CwbTjK&index=32.

final research was on telomeres, the code sequences at the tips of the DNA molecule that are supposed to regulate cell longevity; in cancer cells the telomeres fail to shorten as they should.[11] Vogt moved with Dulbecco when he made the transition to the Salk Institute in 1962, and by the time of her retirement in the late 1990s, she had the distinction of being the longest-working scientist at the institute, having spent more years at a Salk bench than anyone else. No doubt she spent more hours, as well, as her strong work ethic had her put in ten-hour days for six days a week. Sundays were reserved for social affairs at her home, which were attended by her students, visiting professors, and colleagues and always featured music. She ran on the beach and did push-ups and calisthenics until she was eighty-six years old. In addition to being legendary for teaching exact scientific techniques, she was famous for her generosity. She'd often assist struggling students with cash, once buying a student a car to replace his wrecked vehicle, and was fondly remembered for her steadfast encouragement and friendship. Four of her students would go on to win Nobel Prizes. In a 2001 interview with the *New York Times*, Vogt was asked if she minded being passed over for recognitions such as the Nobel Prize. "I'm happy not to have been bothered," she said. "When you get too famous, you stop being able to work."[12] Marguerite Vogt died in 2007 at the age of ninety-four.

Dulbecco returned to the Salk Institute in 1977 and eventually served as its president. He recognized that basic research activities were not likely to make breakthroughs in human cancer treatments for decades to come. He urged governments to continue to support

---

11 M. H. Shen, C. Haggblom, M. Vogt, T. Hunter, K. P. Lu, "Characterization and Cell Cycle Regulation of Related Human Telomeric Proteins Pin2 and TRF1 Suggest a Role in Mitosis," *Proceedings of the National Academy of Sciences of the United States of America* 94 (1998): 13618–23, doi.org/10.1073/pnas.94.25.13618.

12 N. Angier, "Scientist at Work—Marguerite Vogt; A Lifetime Later, Still in Love with the Lab," *New York Times,* April 10, 2001.

research while also focusing on steps to prevent cancer, namely, to control the relentless introduction of noxious chemicals into mainstream use in industry and by consumers and to enact legislation to sharply curtail smoking. At the conclusion of his Nobel lecture, Dulbecco said:

> Historically, science and society have gone separate ways, although society has provided the funds for science to grow and in return science has given society all the material things it enjoys. In recent years, however, the separation between science and society has become excessive, and the consequences are felt especially by biologists. Thus, while we spend our life asking questions about the nature of cancer and ways to prevent or cure it, society merrily produces oncogenic [cancer causing] substances and permeates the environment with them. Society does not seem prepared to accept the sacrifices required for an effective prevention of cancer.[13]

In a 2005 addendum to his Nobel website autobiographical note, Dulbecco said he looked forward to retirement at the age of ninety-two to play the piano.[14] One of his last publications (in 2007) showed that mammary stem cells initiate breast cancer.[15]

Renato Dulbecco died in 2012, just two weeks short of his ninety-eighth birthday.

---

13 R. Dulbecco, "From the Molecular Biology of Oncogenic DNA Viruses to Cancer," December 12, 1975, NobelPrize.org, nobelprize.org/prizes/medicine/1975/dulbecco/lecture/.

14 R. Dulbecco, Biographical, NobelPrize.org. nobelprize.org/prizes/medicine/1975/dulbecco/biographical/.

15 I. Zucchi, S. Sanzone, S. Astigiano, P. Pelucchi, M. Scotti, V. Valsecchi, O. Barbieri, G. Bertoli, A. Albertini, R. A. Reinbold, and R. Dulbecco, "The Properties of a Mammary Gland Cancer Stem Cell," *Proceedings of the National Academy of Sciences of the United States of America* 104, no. 25 (2007): 10476–81.

Howard Martin Temin was a graduate student in Dulbecco's lab at Caltech.

Temin was born in Philadelphia in 1934. His parents were social activists, and Temin carried on this tradition from a young age by donating his bar mitzvah money to a refugee camp. As the high school valedictorian, Temin gave a speech that addressed the potential of science to do great good, such as putting a man on the moon, and to do great harm, such as with the atomic bomb. Temin's first exposure to professional science was through a summer program offered to high school students at Jackson Labs in Bar Harbor, Maine, which was famous for providing most of the world's labs with rats and mice. He also spent one summer at the Institute for Cancer Research in Philadelphia, and published his first scientific paper (as a coauthor) at the age of eighteen.[16]

Temin attended Swarthmore College, where he was known to undertake a great deal of independent study and to challenge his professors. The yearbook tagged him as "one of the future giants in experimental biology." Temin attended graduate school at Caltech, where he worked in the lab of his future cowinner Renato Dulbecco. His significant contribution there was to develop a method for quantifying animal virus particles by observing the shape changes they caused in infected cell cultures.[17] This was analogous to the plaque-counting method to detect viral particles that created holes when they killed off infected bacterial cultures. Temin's method counted cell colonies whose shapes were changed by the viral infection, rather than holes from being killed off. This method allowed for a tremendous expansion in the study of animal viruses in cells.

While still a graduate student at Caltech, Temin made observations that suggested that the Rous sarcoma virus (RSV) becomes

---

16 T. Ingalls, F. Avis, F. Curley, and H. Temin, "Genetic Determinants of Hypoxia-Induced Congenital Anomalies," *Journal of Heredity* 44, no. 5 (September 1953): 185–94.

17 H. Temin and H. Rubin, "Characteristics of an Assay for Rous Sarcoma Virus and Rous Sarcoma Cells in Tissue Culture," *Virology* 6, no. 3 (1958): 669–88.

incorporated as a part of the genome of the host cell, similar to how some bacteria-infecting viruses were known to incorporate into bacterial DNA. But since RSV was an RNA virus, no one knew how this might occur. Up to that the point, the cherished theory was that gene copying only went in one direction, wherein DNA is transcribed by RNA and then RNA makes proteins. A researcher of the era describes the immense pressure to discourage questioning the idea: "It was almost heresy to break with the central dogma of the times, which offered no exceptions to the rule that the information flow was only from DNA to RNA."[18] Yet Temin had no other explanation for his observation that subsequent generations of cells contained the virus. Temin's idea called for the process to work in the opposite direction, with the RNA of the virus getting copied into the DNA of the host cell. This was very unpopular, and Temin drew criticism for years while he steadfastly looked for the mechanism of this reverse action. Temin also proposed a related theory: there existed a "provirus"—some aspect of a virus that existed within the host cell and was inherited by subsequent cell generations, only later emerging as an active and potentially cancer-causing virus.[19]

From Caltech, Temin moved to the University of Wisconsin, where he remained for the rest of his career. Temin initially worked with only two lab technicians in a basement laboratory amid the sump pumps and steam pipes of the biology building, taking on no graduate students for his first few years. He conducted a series of experiments seeking the mechanisms that would validate his theories, never giving up despite many failures. What drove his work was the certainty that this line of inquiry could shed light on the development of human cancers. Although very few human cancers are related to viral infections, the study of cancer-causing viruses in cell cultures could reveal the

---

18 T. W. Mak, "Cherish an Idea that Does not Attach Itself to Anything," in *The DNA Provirus: Howard Temin's Scientific Legacy*, ed. R. G. Temin and B. Sugden (Washington, DC: ASM Press, 1995).

19 H. Temin, "Cancer and Viruses," *Engineering and Science* 23, no. 4 (1960): 21–24.

mechanisms of how healthy cells transform into cancer cells and thus lead to new cancer treatments. His tenacity in pursuing an unpopular theory has been called the Temin effect, attributed to shrewd powers of observation and reasoning along with a strong imagination.

Finally, in 1970, Temin succeeded in finding the mechanism, demonstrating that viral RNA gets turned into DNA by the action of an enzyme that drives the backward reaction.[20] Working independently, his future Nobel cowinner David Baltimore found the same enzyme by an entirely different method. The enzyme was later named reverse transcriptase because it transcribes genetic-code information in the reverse order from the usual. Almost sixty years later an article on thinking outside the box commented, "The discovery of reverse transcriptase was a challenge to a tenet of modern biology so hallowed it was actually called 'the central dogma': genetic information flows from DNA to RNA and from there is encoded in proteins. Temin argued it could also flow in the opposite direction—a claim that was treated with derision. To imagine a possibility so heretical required imagination."[21]

Temin had worked on his unpopular theory for many years, all the while facing intense criticism. In contrast, his cowinner Baltimore had only been seeking the enzyme for a short time. Unlike Baltimore, Temin was neither adversarial nor strident, though he never compromised on what he knew he had observed.

Temin received the Nobel Prize on his forty-first birthday. In his Nobel lecture, Temin acknowledged that his research efforts were funded by the American taxpayers through government grants, mentioning in particular the National Cancer Institute and the American Cancer Society.[22] In another striking difference from his cowinner

---

20 H. Temin and S. Mizutani, "RNA-Dependent DNA Polymerase in Virions of Rous Sarcoma Virus," *Nature* 226 (1970): 1211–13.

21 D. Epstein, M. Gladwell, "The Temin Effect," *Ophthalmology* 125, no. 1 (2018): 2–3.

22 H. Temin, "The DNA Provirus Hypothesis," December 12, 1975, NobelPrize.org, nobelprize.org/prizes/medicine/1975/temin/lecture/.

David Baltimore, Howard Temin took extraordinary pains to avoid benefiting personally from commercial applications related to his work.

Temin took the opportunity of his Nobel banquet speech to decry cigarette smoking in a room where many of the guests were puffing away in their tuxedos and evening gowns. Speaking on behalf of all three of the prizewinners, Temin said, "We are, in fact, outraged that the one major measure available to prevent much cancer, namely, the cessation of cigarette smoking, has not been more widely adopted."[23] His sense of social responsibility as a scientist harkened back to his early public address when he was just a high school valedictorian, apologizing to the Nobel guests for the destructive use of science by the US. He ended the speech by hoping for peace: "Finally, although science can be applied both for constructive and destructive purposes, we hope that science will be in the future, as are these Nobel Prizes, only for peaceful purposes."[24]

Temin's later research contributed to experiments on vaccines for AIDS, which was a natural progression because HIV is also an RNA virus (a "retrovirus"). Temin observed that there are two kinds of RNA retroviruses, simple ones that infect chickens, and complex ones (like HIV) that infect humans. Temin proposed developing a simple retrovirus that could replicate in humans, suggesting that it might provoke an immune response against the more complex HIV.[25] This would not be a vaccine in the traditional sense but would infect recipients with a related virus to develop immunity to the real HIV. Temin also ceaselessly advocated for more funding for AIDS research.

When in his late fifties, Howard Temin developed adenocarcinoma of the lung, which at the time was not thought to be related to

---

23 H. Temin, Banquet Speech, December 10, 1975, NobelPrize.org, nobelprize.org/prizes/medicine/1975/temin/speech/.

24 H. Temin, Banquet Speech, December 10, 1975, NobelPrize.org, nobelprize.org/prizes/medicine/1975/temin/speech/.

25 H. M. Temin, "A Proposal for a New Approach to a Preventive Vaccine against Human Immunodeficiency Virus Type 1," *Proceedings of the National Academy of Sciences of the United States of America* 90, no. 10 (1993): 4419–20.

smoking (he had never smoked). It has since been demonstrated that adenocarcinoma is strongly associated with a history of smoking, but still many nonsmokers can get this kind of cancer.[26] It is not known if Temin had other lung cancer risk factors such as exposure to second-hand smoke, radon, air pollution, or chemicals and materials in his work environment. Howard Temin died in 1994 at the age of fifty-nine. He has been fondly remembered as being scientifically rigorous as well as kind and generous.[27]

———————

David Baltimore was the youngest of the trio sharing the 1975 Nobel Prize. He was born in 1938 in New York. He sought action in research labs as early as his high school years, having been accepted into a summer program at Jackson Labs in Bar Harbor, Maine. There he met future Nobel Prize cowinner Howard Temin. Temin was only four years older, but he had already published his first scientific paper at the age of eighteen, which helped Baltimore to realize that a serious career in science was well within his reach. That summer, Baltimore participated in genetics research—a subject that was not being taught in high school nor even at many colleges.

Baltimore enrolled as a biology major at Temin's alma mater, Swarthmore College, where Temin's reputation was a constant impetus for Baltimore to do better. But Baltimore was disappointed by the biology classes, which involved much memorization and no hands-on work. He switched to a major in chemistry, where there was plenty of laboratory activity. In his sophomore year, Baltimore worked in the labs of Mount Sinai in New York trying to extract likely cancer-fighting substances from sea cucumbers. The next summer he worked at Cold

---

26 P. Yang, J. R. Cerhan, R. A. Vierkant, J. E. Olson, C. M. Vachon, P. J. Limburg, A. S. Parker, K. E. Anderson, and T. A. Sellers, "Adenocarcinoma of the Lung Is Strongly Associated with Cigarette Smoking: Further Evidence from a Prospective Study of Women," *American Journal of Epidemiology* 156, no. 12 (2002): 1114–22.

27 See, R. G. Temin, introduction to *The DNA Provirus: Howard Temin's Scientific Legacy*, ed. T. G. Temin and B. Sugden (Washington, DC: ASM Press, 1995).

Spring Harbor, where he gained experience in handling bacteria and viruses. During his senior year, he worked in the chemistry labs at the University of Pennsylvania extracting the enzymes involved in cellular energy production.

Baltimore next attended MIT, where once again he became impatient listening to professors explain science rather than rolling up his sleeves in the lab. He used connections made at Cold Spring Harbor to get accepted on an MIT project attempting to define the DNA of bacteriophages. This put him on the forefront of biomedical research, because several academic groups around the world were competing to be the first discover the genetic composition of viruses, which were expected to open the door to understanding the human gene. The MIT team made a breakthrough when they found their bacteriophage had a single chromosome consisting of about fifty thousand molecules. (It was eventually found that human DNA is about one hundred thousand times longer.)

Baltimore really enjoyed lab work, calling it "sanctioned self-indulgence,"[28] but by this time he recognized that the bacteriophage field was getting crowded, thereby offering less opportunity for any one researcher to gain distinction. He wanted to strike out into an area where he would be more likely to earn recognition, and he had a hunch that animal virology would be just the thing, thinking that studying viruses that affect animals might also be more applicable to humans. He decided to further his ambitions by transferring to the graduate program at the Rockefeller Institute in New York City, where animal viruses were under study.

At the Rockefeller labs, Baltimore perfected his ability to grow cultures of cells derived from mice and infect them with the mengovirus, a virus that infects rodents and cattle. Mengovirus was found to be an RNA virus. Up to that point it was understood that RNA viruses

---

28 D. Baltimore, "On Doing Science in the Modern World," Tanner Lectures on Human Values, delivered at Clare Hall, Cambridge University, March 9 and 10, 1992, tannerlectures.utah.edu/_documents/a-to-z/b/Baltimore93.pdf.

usurped the infected cells' apparatus to get the host DNA to make more copies of the viral RNA. But Baltimore's first discovery showed that must not be the only way viruses replicate: he found that when mengovirus infected a cell it blocked the cell's DNA code from being transcribed by RNA. This posed a puzzle: Since the virus blocks DNA, how then does an RNA virus get replicated? Baltimore postulated that there must be an enzyme that assists RNA replication, and through a series of experiments, he proved its existence.[29] He further showed that poliovirus was an RNA virus similar to mengovirus, and it too used the enzyme named RNA replicase to make more copies of its own RNA.

Baltimore's animal-virus research was robust enough to submit as a PhD thesis, but he had only been enrolled at Rockefeller for a mere eighteen months—it was unheard of to earn a doctorate degree in such a short time. Yet Baltimore had already been offered a paid postdoctorate position back at MIT, where he aimed to gain more experience with viral enzymes than Rockefeller could offer, and he was determined to move on. The compromise was for Baltimore to take his time writing up his thesis while he was working elsewhere; this he did, and he was awarded his PhD in 1964.

Baltimore worked at MIT for only a few months before moving over to Albert Einstein College to learn from an even more experienced enzymologist. Although it was not precisely on the track of his research, Baltimore decided to find out what chemical reaction triggered the RNA replicase enzyme that some RNA viruses use to make more RNA. In gross violation of an unwritten rule of courtesy in highly competitive labs, Baltimore borrowed some RNA enzyme from someone else's bench and borrowed some radioactively labeled phosphate from another. Using these substances, Baltimore quickly demonstrated that RNA replicase is triggered by a phosphate reaction. This turned out to be valuable information for labs around the world

---

29 D. Baltimore and R. M. Franklin, "Preliminary Data on a Virus-Specific Enzyme System Responsible for the Synthesis of Viral RNA," *Biochemical and Biophysical Research Communications* 9 (1962): 388-92.

who were working on the biochemistry of intracellular machinery. It is a weird commentary on the realities of scientific competition that the researchers whose substances were "borrowed" for Baltimore's experiment immediately claimed they were due some credit—as if they had been just about to do such an experiment themselves—and so when the paper came out it listed Baltimore as the third author.[30] Those who do the actual work are supposed to be given top billing.

Despite his bad laboratory manners, Baltimore was recruited by future Nobel cowinner Renato Dulbecco to join him at the Salk Institute in San Diego for more poliovirus research. At Salk, Baltimore demonstrated that RNA replicase was also in other viruses, but that did not fully explain how the virus used the host cell's machinery to make viral RNA in the first place. It never took long for Baltimore to become disgruntled, but his beefs at the Salk Institute were not so much about the science. With the same belligerence he brought to protests against the Vietnam War, Baltimore thought that since he was such a stellar scientific contributor, he ought to have more of a say in how the joint was run. The seven-member board of trustees did not agree.

Baltimore returned to MIT, where once again he applied his out-of-the-box thinking. He was pondering the ways in which some RNA viruses manage to incorporate their viral genetic material into the host DNA. Within a month, Baltimore discovered the existence of a viral enzyme that reads the viral RNA into the host cell DNA. This action makes viral genetic code part and parcel of the host DNA. From then on, the host DNA generates instructions to make more viral RNA. This discovery shook up molecular biology's "central dogma," which was that RNA always copies the code of DNA, not the reverse. Howard Temin had been working on the same problem for seven years and finally made the identical discovery just before Baltimore had, albeit using an entirely different approach. Their papers were published in

---

30 U. Maitra, A. Novogrodsky, D. Baltimore, and J. Hurwitz, "The Identification of Nucleoside Triphosphate Ends on RNA Formed in the RNA Polymerase Reaction," *Biochemical and Biophysical Research Communications* 18, nos. 5–6 (1965): 801–11.

1970 in the same issue of the journal *Nature*.[31] This was the work that won them the Nobel Prize in 1975.

After his Nobel Prize, Baltimore continued to engage in controversies that were on the forefront of biology. In 1971, when the US president announced the "War on Cancer," Baltimore lobbied against funding for it, fearing that too much money would be wasted on trying useless remedies in patients rather than on valuable basic research to understand the fundamentals of cancer development. He changed his views a couple of years later when he was awarded an American Cancer Society professorship worth $1 million in salary support over thirty years (as long as he continued in cancer research). In 1970, Baltimore took the opportunity of accepting an Eli Lilly award to espouse his anticapitalist views by berating the pharmaceutical company for fixing prices on medicines. He would later become a principal in a biotech company and submit patents for hundreds of products and processes, much of it discovered in publicly-funded research projects. Many do not realize that universities and individual researchers are allowed to own and patent innovations paid for by taxpayer dollars. It was not that way until the Patent and Trademark Law Amendments Act of 1980 (the Bayh-Dole Act) granted research institutions and scientists the ability to own inventions, along with an obligation to exploit and monetize them.[32]

———

The discoveries of Baltimore and Temin and all of the other molecular biologists of the previous twenty-five years combined to make genetic engineering a reality. In 1971, the future Chemistry Nobel

31 D. Baltimore, "RNA-Dependent DNA Polymerase in Virions of RNA Tumour Viruses," *Nature* 226, no. 5252 (1970): 1209–11; H. M. Temin and S. Mizutani, "RNA-Dependent DNA Polymerase in Virions of Rous Sarcoma Virus, *Nature* 226, no. 5252 (1970): 1211-13, doi.org/10.1038/2261211a0, erratum in *Nature* 227, no. 5253 (1970): 102.

32 For the Bayh-Doyle regulations, see the NIH website at grants.nih.gov/grants/bayh-dole.htm.

Prize–winner Paul Berg succeeded in making the first recombined DNA by splicing together some DNA from a normal bacterium with that of a bacteria-infecting virus and an animal virus. The animal virus, SV40, could cause cancer in some species of animals but had not been connected conclusively with any cancers in humans. Yet Berg worked with it carefully, using all of the latest biohazard safety techniques in his lab. It turned out that these measures were useless, as all of his lab personnel were infected with SV40, though with no ill effects. What would be the future effects if there was an escape of a new, lab-created unnatural mix of genetic material combined with SV40's high infectivity? Berg's lab made their monster on purpose, but the following year, some labs found hybrid adenoviruses (which causes the common cold) mixed with SV40 were unexpectedly lurking in their specimens.

While those first man-made genetic freaks did not seem to affect human health, in 1973 Berg and his colleagues successfully spliced genes that caused bacteria to become resistant to antibiotics. Now it was getting creepy — what if antibiotic resistant bacteria spread outside the lab and dominated Earth, killing off large populations?

Laboratory Escapees

In 1967 when twenty-five laboratory workers became ill simultaneously in Marburg, Germany, and Belgrade, Yugoslavia, it was discovered they had all worked with shipments of green monkeys from Uganda. Six additional people became ill from caring for the lab workers. Seven of the lab workers died. They were found to be infected with a virus carried by the monkeys, named Marburg, after the town in Germany. Marburg is in the same family as Ebola virus, which was not discovered until nine years later, when there were deadly human outbreaks in Zaire and Sudan, although in those outbreaks, the victims had no contact with monkeys.

In 1989, dozens of macaques imported from the Philippines were received in Reston, Virginia, by a private company that sells lab animals. While they were in quarantine, dozens of the animals suddenly died of hemorrhagic fever. Initial testing showed the Ebola virus Zaire strain, which had a 90 percent human fatality rate. Four lab workers were found to be infected. An army unit was mobilized to kill the remining animals and disinfect the building. The four humans did not even fall ill, and later testing showed they had gotten a rare strain, now called Ebola-Reston, the only one of the five Ebola strains not fatal to humans.[33] Ebola-Reston reappeared in 1996, when two monkeys imported from the Philippines to the same company's facility in Alice, Texas, died. Two animal handlers at the facility were infected, despite attesting to using all of the recommended safety gear and following strict protocols. They did not become ill.[34]

Subsequent events involving Ebola virus were near misses. In 2018, a visiting Chinese researcher at Canada's National Microbiology Lab in Winnipeg arranged for Ebola virus and other pathogens to be shipped to Wuhan. According to a report by the Canadian Broadcasting Company in 2019, heavily redacted emails reveal a back-and-forth discussion in which inadequate packaging material was questioned by a person in the receiving Chinese lab, who asked, "Aren't you making a mistake here?" This alone prompted a correction of the packaging and handling, including disinfection of the outside of the package before it was loaded onto an Air Canada flight.[35]

In 2019, the US government issued a cease-and-desist order to the US Army's bioweapons facility at Fort Detrick, Maryland. The

---

33 "Ebola Virus Infection in Imported Primates—Virginia, 1989," *MMWR Weekly* 38, 48 (December 8, 1989): 831–32, 837–38.

34 "Ebola-Reston Virus Infection among Quarantined Nonhuman Primates—Texas, 1996," *MMWR Weekly* 45, no. 15 (April 19, 1996): 314–16.

35 K. Pauls, "Canadian Scientist Sent Deadly Viruses to Wuhan Lab Months before RCMP Asked to Investigate," CBC News, June 14, 2020, cbc.ca/news/canada/manitoba/canadian-scientist-sent-deadly-viruses-to-wuhan-lab-months-before-rcmp-asked-to-investigate-1.5609582.

lab was handling Ebola, smallpox, and anthrax. They had previously used steam sterilization to treat laboratory wastewater, but when a flood affected their lab in 2018, they switched to chemical sterilization. Inspectors from the CDC found the chemical processes to be inadequate, saying it potentially allowed for discharge of infected material into public waterways.[36]

This prompted David Baltimore to join with Paul Berg, Jim Watson, and other leading researchers to hold a biohazard conference, issue a book on the subject, and call for a worldwide moratorium on genetic engineering until safeguards could be established. In 1975, they hashed out safety guidelines that were readily embraced by the National Institutes of Health and became a standard around the world. This did little to appease the mayor of Cambridge, Massachusetts, where MIT is located, who issued a local moratorium on gene research. It generated a very public debate about "bad genes" and "good genes" and whether mad scientists should really be allowed to police themselves. It was finally resolved by banning research only on things that could become bioterrorism weapons. Research activity in genetic engineering subsequently exploded, and today many vaccines, drugs, hormones, foods, and textile crops are the product of genetic engineering. The initially tight restrictions relaxed over the next two decades.

In 1986, Thereza Imanishi-Kari, an immunologist Baltimore had previously collaborated with, joined MIT. That same year, Baltimore jointly published a paper with her and two others at MIT in the journal *Cell*. Baltimore was listed as third author, but apparently his major contribution was only to suggest the direction of research (on antibody responses in tumors found in mice). Trouble started when Margot O'Toole, one of the postdoctoral students under Imanishi-Kari, raised

36 T. Wyatt, "Research into Deadly Viruses and Biological Weapons at US Army Lab Shut Down over Fears They Could Escape," *Independent*, August 6, 2019, independent.co.uk/news/world/americas/virus-biological-us-army-weapons-fort-detrick-leak-ebola-anthrax-smallpox-ricin-a9042641.html.

concerns about the validity of her boss's data in the paper. O'Toole found lab notes that suggested to her that at least one of the mice used in the experiment had been incorrectly categorized. This was no smoking gun, as Imanishi-Kari had recognized the error and already accounted for it in the experiment. Later, O'Toole found evidence that the reagents used in the experiment were incorrectly described. While this was true, it was a minor error that did not make or break the results. Such an error would normally be handled by O'Toole writing up the discrepancy to the editor of the journal, as indeed Baltimore urged her to do. Alternately, any of the authors could have written up the minor correction. It certainly was not fraud. But neither O'Toole nor Baltimore nor any of the paper's other authors submitted the correction to the journal. O'Toole persisted in her accusations of "fraudulent research," finally prompting Baltimore to ask for an NIH inquiry into the paper.

The inquiry erupted into a national debate involving a congressman, media interviews, federal hearings, and the involvement of the Secret Service to do forensic analysis on Imanishi-Kari's lab notebooks. It was complicated terribly by the hard feelings between O'Toole and Imanishi-Kari, the latter's imperfect English, and vicious rumors fueled by the congressman's aides. David Baltimore's angry public tirades, in defense of Imanishi-Kari in particular and in defense of the right of scientists to police themselves in general, did not contribute to a calm resolution. His cogent and rational reasoning got lost in the emotion of it all.

Baltimore also crafted a written submission to the federal hearings on the matter, which appeared in the *Federal Register*. He made a major point that the simple test of the validity of a scientist's work is the reproduction of it by independent researchers. Some of his colleagues had urged Baltimore to reproduce the questionable experiment himself; although not quite the same as if it had been done by outsiders, this would still go a long way to defending the original paper. He declined to do so on the basis that he was not an immunologist, which raises the question of what his name was doing on the paper if he was not qualified to do the research and could not authoritatively vouch for its veracity?

Meanwhile, Baltimore accepted a new position as the president of Rockefeller University. He immediately made major changes, including salary structures and promotions, which put him at odds with senior faculty. A year later, the federal investigation concluded that Imanishi-Kari had committed fraud, based largely on a Secret Service report of suspicious aspects of her lab notebooks. This finally threw Baltimore over, and he abruptly switched his allegiance. Baltimore issued a lengthy apology admitting that he should never have defended Imanishi-Kari, and he asked the journal to retract the paper. This was a victory for O'Toole, who rushed to the media with fresh descriptions of the supposed deceptions by Imanishi-Kari and others allegedly involved along the way. O'Toole's new accusations included Baltimore and many others who had been sympathetic and had in fact tried to aid her cause. This set Baltimore off anew, with yet another switch to defending the paper he'd just retracted. He used his considerable status to get letters published in journals about O'Toole's twisting of the facts.

Baltimore's enmeshment in the ongoing drama—first with a strident defense of the accused, then the mea culpa letter, followed by the fresh retort to O'Toole—was a bit much for the Rockefeller seniors, who were already struggling with Baltimore's leadership. By mutual agreement, Baltimore resigned. Imanishi-Kari immediately sought an appeal, and this time the investigation was thorough and markedly less political. Among other things, the new investigation included an outside lab analysis of the Secret Service report, and found it to be rubbish. Two years, later Imanishi-Kari was quietly fully exonerated of fraud.

Retracted COVID Papers

In 2020, the first year of the pandemic, RetractionWatch.com listed a total of sixty-eight papers on COVID that were retracted.[37]

In 1995, Baltimore became president of the California Institute of Technology, where he remained until retirement and then continued as professor emeritus. Another research controversy dogged him when a previous postdoctoral student confessed to fraud. In 2004, Luk van Parijs was an associate professor at MIT when he was confronted by lab associates with evidence that he had falsified data.[38] This opened up investigations into any work he had ever submitted, including the two papers coauthored with Baltimore years earlier when van Parijs was a postdoctoral student at Caltech.[39] Those papers were found to include fraudulent data and have since been retracted. There were no repercussions for Baltimore, who, though given last author billing, was the senior figure on the papers. Furthermore, Baltimore and one of van Parijs's former supervising professors at MIT successfully pleaded to the court for sentencing leniency for van Parijs, who escaped a jail term and instead was sentenced to house arrest, community service, and being banned from receiving federal research grants for five years.[40] This time, there was surprisingly little outcry about scientists policing themselves.

---

37 "Retracted Coronavirus (COVID-19) Papers," Retraction Watch, retractionwatch.com/retracted-coronavirus-covid-19-papers/.

38 "M.I.T. Dismisses a Researcher, Saying He Fabricated Some Data," *New York Times*, October 28, 2005.

39 L. van Parijs, Y. Refaeli, J. D. Lord, B. H. Nelson, A. K. Abbas, and D. Baltimore, "Uncoupling IL-2 Signals That Regulate T Cell Proliferation, Survival, and Fas-Mediated Activation-Induced Cell Death," *Immunity* 11, no. 3 (1999): 281–88; retraction in L. Van Parijs, Y. Refaeli, J. D. Lord, B. H. Nelson, A.K. Abbas, D. Baltimore, *Immunity* 30, no. 4 (2009): 611; L. van Parijs, Y. Refaeli, A. K. Abbas, and D. Baltimore, "Autoimmunity as a Consequence of Retrovirus-Mediated Expression of C-FLIP in Lymphocytes," *Immunity* 11, no. 6 (1999): 763–70, retraction in L. van Parijs, R. Refaeli, A. K. Abbas, and D. Baltimore, *Immunity* 30, no. 4 (2009): 612.

40 The outcome of van Parij's federal investigation and sentencing is found in *Federal Register* 4, no. 14, Notices: 4201–2.

Baltimore's biographer relates quotes from roommates and professors and colleagues, variously describing David Baltimore as arrogant, an exceptional and decent human being, and a pain in the ass.[41] He certainly made significant contributions to science.

Baltimore dispelled myths about how science is conducted: "Any pursuit so far beyond present knowledge to be pathfinding is not a logical progression from the established order. That, in fact, is the dirty secret of high science—it is not logical. At best it is analogical, and often one is driven simply by a hunch."[42]

---

41 S. Crotty, *Ahead of the Curve: David Baltimore's Life in Science* (Berkeley: University of California Press, 2001).

42 D. Baltimore, "Discovery of the Reverse Transcriptase," *FASEB Journal* 9 (1995): 1660–63.

Afterword

What comes with winning the Nobel Prize?

There's the gold—solid 23-karat before 1980 but since then made of an alloy plated with 24-karat gold. Then there's the cash, with the first prize in 1901 giving winnings of 150,782 Swedish kronor (SEK), valued in 2020 currency at SEK 8,722,510. After 1901, the cash prize decreased for many years, depending on the performance of the invested funds, hitting a low in 1919 of only 28 percent of the original amount and staying below 40 percent until 1974. The prize money started a consistent climb in 1983 and has been nearly at, or exceeding, 100 percent of the original amount since 1991. The prize was set at SEK 10,000,000 in 2020 (approximately 1.17 million USD). The money is shared more often than not, with only six prizes in this quarter-century awarded to single individuals. Sometimes it was not shared equally. For example, in 1958 half the prize went to Joshua Lederberg while the other half was split between George Beadle and Edward Tatum.

The Nobel Prize garners prestige, particularly scientific and academic prestige, which facilitated some winners to become administrators of illustrious institutions, like James Watson at Cold Spring Harbor and David Baltimore at Caltech. For others, the prestige gained them comfortable positions conducting dead-end research for the rest of their lives, such as Francis Crick, who didn't manage any significant contributions in his twenty-eight years at the Salk Institute. In contrast, after winning his Nobel Prize, Har Gobind Khorana used his new position at MIT to lead the team that created the first man-made gene, and through the remainder of his career, he continued to contribute significantly to the understanding of gene control.

The tendency for authority to be assigned to Nobel laureates has been seized upon by many prizewinners to make pronouncements and offer opinions on subjects far afield of their training and expertise. After making tremendous contributions in the fields of virology and immunology, Frank Burnet served as a consultant on Australian educational policies with recommendations to do away with courses in literature, history, and art. Nikolaas Tinbergen went from studying dogs and parakeets to setting himself up as an autism expert, primarily targeting diffident mothering. His cowinner Konrad Lorenz concentrated his research on the behaviors of geese but felt entitled to declare with authority that Vietnam War protestors were suffering from mass mental illness. Perhaps most egregious of this lot was James Watson, who relentlessly claimed a genetic basis for everything from stupidity and laziness to sexual appeal.

Others embraced their authority to resort to espousing esoteric philosophies, with a surprising number tackling the nature of consciousness. Christian de Duve considered that the spiritual and physical aspects were one, while Gerald Edelman was rather more specific in explaining that consciousness was the very stuff of brain connections. In contrast, John Eccles postulated an intelligence beyond the body that was controlling it. Ragnar Granit admitted that nothing in his deep study of brain neurons shed light on consciousness and predicted a dead end for researchers who would try to add up the sum of mechanical parts to explain spiritual awareness. George Wald concluded that consciousness has no location, cannot be measured, and is independent of space and time. Another favorite topic of these laureate/philosophers was to speculate on the origin of life. The two camps were sharply divided: either life emerged after a series of incredibly lucky accidents, as espoused by George Beadle, George Edelman, and Jacques Monod; or there was some intelligence or higher power involved, as speculated by John Eccles and Albert Claude. Then again it could be aliens, as outlined in Francis Crick's theory of panspermia.

Many of these winners recognized that the fame, prestige, and authority of being a Nobel laureate also confers some social responsibility, but a special few of these Nobel Prize winners were heroes long before conducting their prizewinning research. Albert Claude risked his life to work underground with British intelligence in Belgium during WWI. The French freedom fighters André Lwoff, François Jacob, and Jacques Monod narrowly escaped death during WWII. Howard Temin was speaking against nuclear war as early as high school.

Salvador Luria protested the Vietnam War, nuclear proliferation, and inequality in employment and health care. Maurice Wilkins separated radioactive plutonium in government labs in England and then joined the Manhattan Project in the US, but later became an anti-nuclear activist. Marshall Nirenberg fought political oppression and the proliferation of nuclear arms. David Baltimore engaged in Vietnam War protests. The most eloquent of these men was George Wald, who challenged the public to look deeper into the institutions and policies that were driving the planet ever closer to nuclear annihilation.

It is my hope that these stories convey an appreciation for the strengths and frailties of the people who make up the winners of the highest medical recognition in the world. If it helps you cultivate a healthy skepticism when you hear pronouncements from the medical experts of the day, I will feel I have communicated well.

Look for the next volume in the Boneheads and Brainiacs series in the near future.

Winners of the Nobel Prize in Physiology or Medicine 1951–1975

1951   Max Theiler "for his discoveries concerning yellow fever and how to combat it"[1]

1952   Selman Waksman "for his discovery of streptomycin, the first antibiotic effective against tuberculosis"

1953   Hans Adolf Krebs "for his discovery of the citric acid cycle" Fritz Albert Lipmann "for his discovery of co-enzyme A and its importance for intermediary metabolism"

1954   John Enders, Thomas Weller, and Frederick Robbins "for their discovery of the ability of poliomyelitis viruses to grow in cultures of various types of tissue"

1955   Axel Hugo Theodor Theorell "for his discoveries concerning the nature and mode of action of oxidation enzymes"

1956   André Cournand, Werner Forssmann, and Dickinson Richards "for their discoveries concerning heart catheterization and pathological changes in the circulatory system"

1957   Daniel Bovet "for his discoveries relating to synthetic compounds that inhibit the action of certain body substances, and especially their action on the vascular system and the skeletal muscles"

---

1   All quotes are per the official statements of the reasons for the awards on the Nobel organization official website, "All Nobel Prizes in Physiology or Medicine," NobelPrize.org, nobelprize.org/prizes/lists/all-nobel-laureates-in-physiology-or -medicine/.

1958	George Beadle and Edward Tatum "for their discovery that genes act by regulating definite chemical events" Joshua Lederberg "for his discoveries concerning genetic recombination and the organization of the genetic material of bacteria"
1959	Severo Ochoa and Arthur Kornberg "for their discovery of the mechanisms in the biological synthesis of ribonucleic acid and deoxyribonucleic acid"
1960	Frank Burnet and Peter Medawar "for discovery of acquired immunological tolerance"
1961	Georg von Békésy "for his discoveries of the physical mechanism of stimulation within the cochlea"
1962	Francis Crick, James Watson, and Maurice Wilkins "for their discoveries concerning the molecular structure of nucleic acids and its significance for information transfer in living material"
1963	John Carew Eccles, Alan Hodgkin, and Andrew Huxley "for their discoveries concerning the ionic mechanisms involved in excitation and inhibition in the peripheral and central portions of the nerve cell membrane"
1964	Konrad Bloch and Feodor Lynen "for their discoveries concerning the mechanism and regulation of the cholesterol and fatty acid metabolism"
1965	François Jacob, André Lwoff, and Jacques Monod "for their discoveries concerning genetic control of enzyme and virus synthesis"
1966	Peyton Rous "for his discovery of tumour-inducing viruses" Charles Huggins "for his discoveries concerning hormonal treatment of prostatic cancer"

1967   Ragnar Granit, Haldan Keffer Hartline, and George Wald "for their discoveries concerning the primary physiological and chemical visual processes in the eye"

1968   Robert Holley, Har Gobind Khorana, and Marshall Nirenberg "for their interpretation of the genetic code and its function in protein synthesis"

1969   Max Delbrück, Alfred Hershey, and Salvador Luria "for their discoveries concerning the replication mechanism and the genetic structure of viruses"

1970   Bernard Katz, Ulf von Euler, and Julius Axelrod "for their discoveries concerning the humoral transmittors in the nerve terminals and the mechanism for their storage, release and inactivation"

1971   Earl Sutherland Jr. "for his discoveries concerning the mechanisms of the action of hormones"

1972   Gerald Edelman and Rodney Porter "for their discoveries concerning the chemical structure of antibodies"

1973   Karl von Frisch, Konrad Lorenz, and Nikolaas Tinbergen "for their discoveries concerning organization and elicitation of individual and social behaviour patterns"

1974   Albert Claude, Christian de Duve, and George Palade "for their discoveries concerning the structural and functional organization of the cell"

1975   David Baltimore, Renato Dulbecco, and Howard Temin" for their discoveries concerning the interaction between tumour viruses and the genetic material of the cell"

Index

abortion, 115–116

Academic Assistance Council (AAC), 23, 207

a-cellular pertussis vaccine, 37

acetaminophen (Tylenol), 209–210

acetyl coenzyme A, 136

action potential, 127

adaptation, 103

adenocarcinoma, 275–276

adenoviruses, 281

adrenaline (epinephrine), 217

Adrian, Edgar Douglas, 120

Africa, 2, 8, 281

African Americans, experimentation regarding, 57

aggression, 243, 247

agnosticism, 130

AIDS, 16, 275

Alexander, Frederick Matthias, 248

Alexander technique, 247–248

Amanita, 133–134

American Cancer Society, 274

American Philosophical Society, 24

amino acids, 68, 78–79, 178, 218

ammonia, 22

Anderson, Clarence W. "Herk," 68

anesthesia, 62

angel mushrooms, 134

animals

behavior of, 236, 241–242, 246, 247

cancer research on, 155–156, 160, 162, 163, 254

cell-growth techniques within, 267

experimentation on, 55

hearing range of, 102–103

heart experimentation on, 48

Karl von Frisch and, 227

LSD research on, 212

prion disease of, 155

research regarding, 18, 92, 169, 253

viruses within, 277–278, 281

See also specific animals

antibiotics, 15, 17, 281

antibodies, 92, 221–226

antigens, 92, 221

antihistamines, 63–64

anti-Semitism, 117, 195, 206, 230–231

aqua bionda, 139–140

Architects and Engineers for Social Responsibility, 110

Arrhenius, Svante, 113, 202

astrobiology, 73, 74

atherosclerosis, 141

Atlantic squid, 126–127

atomic bombs, 108–109

atoms, 68

attenuation, 9

Australia, 45, 88, 91–92, 93, 94

autism, 247–248

autophagy, 256

avian leucosis virus (ALV), 11–12, 162

Axelrod, Julius, 202, 209–215

azithromycin, 257

bacteria

antibiotic resistance of, 281

cholera, 154

Coxsiella burnetii (Q fever), 88

DNA and, 145, 199

E. coli, 193

enzymes and, 147–148, 197

formation of, 68–69

genetics and, 70

gonorrhea, 15

gram-negative, 15, 17

group A stretococcus, 154

hand, foot, and mouth disease, 39

H. pylori, 51

overgrowth of, 143

photoreactivation of, 192

pneumococcus, 31

research regarding, 193

Rickettsia, 31

Streptomyces griseus, 18

typhoid, 154

typhus, 30, 50

viruses and, 142, 145, 190

See also tuberculosis (TB)

bacterial antigens, 221

bacteriophages (phages)

discoveries regarding, 267

DNA of, 277

function of, 196–197

photoreactivation of, 192

research regarding, 145, 150–151, 190–194, 195–198, 199

Bailey, F. Lee, 63

Baltimore, David, 265, 274, 276–280, 283–285, 286–287

Bang, O., 155, 162

Banting, Frederick, 50

Barbados, 2

barbiturates, 210–211

bases of DNA, 105

Bayh-Dole Act (Patent and Trademark Law Amendments Act), 280

Beadle, George, 67–70, 75–76

Beadle, Muriel, 68

beer, 133

bees, 227–248

behaviors

of animals, 236, 241–242, 246, 247

innate *versus* noninnate, 243

instinctual, 246

Beijerinck, Martinus, 146

Belgium, 254

Bellevue Hospital, 57

Bender, Lauretta, 57

Berg, Paul, 281, 283

Bergdolt, Ernst, 230

Berger, Gaston, 56

Berkefeld filter, 6

Bernstein, Julius, 131

Beurling, Arne, 205

Beveridge, William, 23

biochemistry, 69

bird flu virus, 90–91

birds, 11–12, 238–239

bladder infections, 15

Bloch, Konrad, 133, 136, 138–140, 141

blood

cancer and, 161

catheterization and, 55

disease transmission through, 154

drug detection within, 63

erections and, 203

leukemia and, 155

liver and, 47

pressure, 204

sampling of, 34

storage of, 157

sugar, 57

typing of, 157

viral multiplication through, 7, 8

yellow fever and, 1, 5–6, 7, 8, 9

Bonner, Yelena, 176

Bovet, Daniel, 60–66

Bovet-Nitti, Filomena, 61, 65

bowel infections, 15

Brabant, Malcolm, 13

brain, 2, 10, 167–171, 225

Brazil, 2

bread molds, 70

breast cancer, 163–164, 269

Briand, Aristide, 26

British Society for Social Responsibility in Science, 110

Brunei, 45

Burkitt's lymphoma, 160

Burnet, Frank, 87–95

caffeine, 210

calcium, 218

Cambridge University, 23, 24

Campa, Cumpersino, 7

Campaign for Nuclear Disarmament, 110

Camus, Albert, 152

cancer

animal research regarding, 155–156, 160, 163

avian leukosis virus and, 11–12

cell growth and, 268–269

chemotherapy on, 163

DNA and, 158

Epstein-Barr Virus (EBV) and, 160–161

genetic-mutation theory of cancer causation, 158

research regarding, 155–165

sarcoma virus, 250–252

self-experimentation regarding, 51

tobacco and, 65

viral DNA and, 268

virus associations with, 160–162, 273–274

cardiopulmonary system, 56

Carro, Antonio, 7

Carroll, James, 5–6, 7

catheters, heart, 47–59

cats, 212, 242–243, 246

cell lines, 89–90

cell respiration, 43

cells

biology of, 252

centrifugation method of, 256

crushing process of, 251

discoveries regarding, 251–252

growth techniques regarding, 267–268

molecular energy currency for, 28

cell/virus system, 146

centrifugation method, 256

Chain, Ernst Boris, 15n2

Charité Hospital, 49–50

Charleston, South Carolina, 2

Charles VI, 134

Chase, Martha, 200–201

chemical energy, 27

chemotherapy, 163

chicken pox *(varicella)*, 31, 39, 88

chickens, 18, 31, 155–156, 160, 162

children, experimentation regarding, 57

Chile, 45

chloroquine, 257

cholera, 154

cholesterol, 133–141

Civilized Man's Eight Deadly Sins (Lorenz), 244

Claude, Albert, 249–253, 263

clonal selection theory, 92–93

cochlea, 98

co-enzyme A, 27

coenzymes, 42, 43

color blindness, 228

communication of bees, 228–229, 233–234

consciousness, 125, 225–226

Continuation War, 167

Coppolino, Carl Anthony, 62–63

Cori, Carl, 78, 216, 219–220, 255, 267

Cori, Gerty, 78, 216, 219, 255, 267

corpses, 97, 154–155

Council for Assisting Refugee Academics (CARA), 25

Council for At-Risk Academics, 25

Cournand, André, 47, 55–56

COVID-19, 2, 38, 155, 257, 286

cows, 155

coxsackie infection, 39

Coxsackieviruses, 39

Coxsiella burnetii (Q fever), 88

crabs, 169

Crawford, Samuel "Wahoo Sam," 68

Creutzfeldt-Jakob disease (CJD), 155

Crick, Francis, 26, 105, 106, 107–108, 111–114, 180–181, 225

Cronstedt, Carola Adelaide, 205

crop science, 77

Cryptocarya alba, 233

Cuba, 4

curare, 61

Cutter Pharmaceuticals, 32

cyclic AMP (cAMP), 217–218

Dale, Henry, 122, 202, 208

Dali, Salvador, 81–82

Darwin, Charles, 129–130

death cap mushrooms, 133–134

deCODE Genetics, 118

de Duve, Christian, 249, 253–261

de Gaulle, Charles, 149

Delbrück, Max, 187–194, 196–197, 199, 267

Delbrück scatter, 188

Delysid, 211–213

DENDRAL, 74–75

deoxyribonucleic acid (DNA)

 bacteria and, 145, 199

 of bacteriophages (phages), 277

 bases of, 105

cancer and, 158

enzyme of, 83–84

function of, 77

generational changes of, 193–194

genetic code within, 180–182

hierarchy of genes within, 151

modification of, 197

outside influences to, 193–194

phage labeling within, 200

recombined, 281

regulatory proteins within, 151

research regarding, 68–69

ribosomes and, 263

RNA and, 86, 178, 274, 279

secrecy regarding, 180

structure of, 105, 106–107, 111–114

ultraviolet light and, 192

viruses and, 84, 268

See also genes

depression, 214–215

d'Hérelle, Félix, 142

diethylstilbestrol (DES), 193–194

digestion, 256

directed panspermia theory, 113

dirt composition, 14–15

Ditzen, Gerda, 48

DNA polymerase, 83–84

DNA virus, 84

dogs, 163, 246, 253

Domagk, Gerard, 50–51

domestication, dangers of, 239

dominators, 167

The Double Helix (Watson), 115

Dozentenakedemie (indoctrination camp), 188

Drucker, Peter, 189

drugs
  acetaminophen (Tylenol), 209–210
  anesthesia, 62
  antibiotics, 15, 17, 281
  azithromycin, 257
  barbiturates, 210–211
  chloroquine, 257
  Delysid, 211–213
  ethambutol, 21
  fomoterol, 257
  hydroxychloroquine, 257
  isoniazid, 21
  lysergic acid diethylamide (LSD), 57, 211–213
  meclizine, 257
  morphine, 210–211
  painkillers, 209–210
  penicillin, 15, 15n2
  Prozac, 213
  pyrazinamide, 21
  quinine, 257
  rifampin, 21
  selective serotonin reuptake inhibitors (SSRIs), 213, 214–215
  streptomycin, 14, 18–20, 21
  succinylcholine (sux), 61–63
  sulfa antibiotic precursor, 50–51
  Thorazine (chlorpromazine), 57
  Wellbutrin, 213
Dulbecco, Renato, 265–271, 279
Duschesne, Ernest, 15n2

ear, 97–104
Eaton, Cyrus, 110
Ebola-Reston, 282
Ebola virus, 154–155, 282
Eccles, John, 120–125, 208
Eccles, Rosamund Margaret, 123
E. coli bacteria, 193

Edelman, Gerald, 221, 223–226
Edmonston, David, 33
Edmonston-Moraten vaccine strain, 35
efficacy, effectiveness versus, 36
Einstein, Albert, 110, 218
Electronics and Computing for Peace, 110
electron microscope, 251–252
electrons, 68
electroshock, 57
Eli Lilly, 85
Ellerman, V., 155, 162
Emergency Association of the German Sciences, 23
Enders, John Franklin, 29, 30–31, 33, 37–38
Enders-Weller-Robbins method, 31–32
Endo, Akira, 140
energy, cell conversion and, 24
energy-providing molecules (ATP), 24, 27–28
Engel, Elsbet, 51
enzymes
  bacteria and, 147–148, 197
  cAMP and, 218
  defined, 41
  discovery of, 178
  DNA, 83–84
  drug exposure and, 210–211
  gene codes for, 70
  light-activated, 192
  liver, 210
  within lysosomes, 256
  Nosema disease and, 231
  research regarding, 78–79
  reverse transcriptase, 274
  RNA, 278–279
  stimulants and, 210
  vitamins and, 151
  See also specific enzymes

epinephrine (adrenaline), 217
Epstein-Barr Virus (EBV), 160–161
Erdos, Tamas, 149
Esperanto, 60–61
espionage, 108–109
estrogen, 163, 164, 193–194
ethambutol, 21
ethology, 241–242
eugenics, 239, 240
evolution, 129–130
exobiology, 73, 74
eyes, 167–172

Farber, Marjorie, 62–63
Farber, William E., 63
Farquhar, Marilyn, 264
fatty acids, 136
Feldman, William, 17
Fenner, Frank, 92
Fermi, Enrico, 196
Fernald State School (Massachusetts), 35
Feynman, Richard, 26
filterable particles, 147, 156, 160
Finlay, Carlos, 3–4, 8
Finsen, Niels, 50
Fischer, Hans, 138
Fleming, Alexander, 15n2
Flexner, Simon, 250, 251
Florey, Howard Walter, 15n2
flouridation of water, 43–45
flu vaccine, 36
fomoterol, 257
Food and Disarmament International, 110
formaldehyde, 32
Forssmann, Werner, 47–55
fovea, 172
France, 142, 143, 144, 148, 149, 151–152
Franchise and Ballot Act of 1892, 121
Francis Crick Institute, 26

Frankfurt University, 189

Franklin, Rosalind, 106, 114, 115

Free French Zone, 144, 148

French Forces of the Interior, 148–149

French vaccine, 10

Friedlander, H., 240

Fromme, Albert, 52–53

fruit flies, 69–70

Fuchs, Klaus, 109

Führer, Wilhelm, 230

Fynen, Feodor, 27

Gage, Matilda Joslyn, 81

geese, 238–239

genes
  appearance and, 69
  bacteria and, 67, 70, 71–72
  carcinogens and, 159
  hierarchy of, 151
  regulatory, 150–151
  research regulations regarding, 283
  See also deoxyribonucleic acid (DNA)

gene therapy, 78

genetically-based bonding, 93–94

genetic decay, 244

genetic engineering, 77, 84–85

genetic-mutation theory of cancer causation, 158

genetic testing, 115–116

genome, 116

gentle gradient technique, 247–248

Germany, 144, 205, 230, 241, 249–250, 262, 281. *See also* Nazis; World War II

germ theory, 143

Gibson, Mary, 63

Gleichschaltung, 230

Glen Grey Act of 1894, 121

glucagon, 217

glucose, 147–148

glucose metabolism, 253–254

Goering, Hermann, 254

gonorrhea, 15

Gorbachev, Mikhail, 176

Gosling, Raymond, 106, 108

gram-negative bacteria, 15, 17

Granit, Ragnar, 166–169

group A stretococcus, 154

Grunberg-Manago, Marianne, 78–80

Guyana, 45

H1N1, 89

H5N1, 90–91

Hammarsten, Einar, 45

hand, foot, and mouth disease, 39

Hanson, Howard, 68

Hartline, Haldan Keffer, 166, 169–171

Harvard University, 35, 91, 115, 173, 198

Havana Yellow Fever Commission, 4, 7

hearing, 97–104

heart, 47–59, 218

Henry VIII, 128

hepatitis A virus, 154

hepatitis B virus, 11, 154, 160

hepatitis C virus, 154, 160

Hershey, Alfred, 187, 191, 198–201

Hertz (Hz), 102

Hess, Rudolf, 189

Heymans, Cornielle, 202

higher education, female employment within, 72

Hill, Archibald Vivian ("A.V."), 202, 206–207

Himmler, Heinrich, 52, 240

Hinshaw, H. Corwin, 17

histamines, 63–64

Hitler, Adolf, 189, 190, 206, 240, 266

Ho Chi Minh, 16

Hodgkin, Alan Lloyd, 26, 120, 126–128, 159

Hodgkin's lymphoma, 160

Holley, Robert, 178, 184–186

Holweck, Fernand, 195, 196

homosexuality, 115

Hong Kong, 45

hormones
  diethylstilbestrol (DES), 193–194
  epinephrine (adrenaline), 217
  estrogen, 163, 164, 193–194
  glucagon, 217
  insulin, 50, 57, 253–254
  noradrenaline/ norepinephrine, 203–204
  prostaglandin, 203
  research regarding, 163, 164, 165
  sex hormones, 165
  substance P, 203
  testosterone, 163

horseshoe crab, 169

Houssay, Bernardo, 202

H. pylori bacteria, 51

Hubbard, Ruth, 173

Huggins, Charles, 153, 162–165

human experimentation, 4–7, 35, 48, 57, 212–213. *See also* self-experimentation

Human Genome Project, 116, 269

human immunodeficiency virus (HIV), 154, 161, 275

human papillomavirus (HPV), 161

Human T-cell lymphotrophic virus type 1 (HTLV-1), 161

Huxley, Aldous, 130

Huxley, Andrew, 120, 126, 127, 128–132
Huxley, Julian S., 130–131
Huxley, Thomas Henry, 129–130
hydrogen, 68
hydrogen peroxide, 140
hydroxychloroquine, 257

ICOS Laboratory, 85
Imanishi-Kari, Thereza, 283–285
immune system, 87, 92
immunization, 37
inducer theory, 242
infanticide, 94
Infected Clothing Building, 5
infectious diseases, from corpses, 154–155. *See also specific diseases*; viruses
Infeld, Russell, 110
influenza, 36, 91–92
influenza A, 31
informed consent, 4–5, 35
innate behavior, 243
instinctual behaviors, 246
Institute of Medicine (IOM), 40
insulin, 50, 57, 253–254
invirent, 238
Irish Republic, 45
isoniazid, 21
Israel, 45

Jacob, François, 142, 149–151, 152
Jaensch, Erich Rudolf, 238
Jensen, Elwood, 164
Jews
    classification of, 238
    exclusion of, 22, 52, 189, 230–231
    labor of, 266
    murder of, 53, 262, 266
    within the Netherlands, 247

refuge for, 205
    segregation of, 230
Jones, Tom, 21
J type personality, 238

Kaiser Wilhelm Institute for Biology, 26
Kaliningrad, 26
Kamprad, Ingvar, 205
Kaposi's sarcoma-associated herpesvirus, 161
Katz, Bernard, 202, 206–209
Katz, Samuel, 33
Keller, Elizabeth B., 184, 185
Khorana, Har Gobind, 178, 183–184
kidney infections, 15
kidneys, 218, 256
King, Martin Luther, Jr., 152
Koch, Robert, 3
Kornberg, Arthur, 77, 83–86, 216–217
Kornberg, Roger, 85–86
Kornberg, Thomas, 84
Kornberg Levy, Sylvy Ruth, 83–84, 85
Krebs, Edwin, 217
Krebs, Hans, 22–23, 24, 25, 26
Krebs cycle, 24, 27
Kuznets, Simon, 26

laboratory-acquired infections, 91
language, development of, 168
The Language of Life (Beadle and Beadle), 68
Laval, Pierre, 144
Laveran, Charles, 4
Law for the Restoration of the Professional Civil Service (Germany), 22, 230
laxness, 58
Lazear, Jesse, 6
Lederberg, Joshua, 70–75
Lederman, Leon, 26

Leigh, Vivian, 16
leiomyosarcoma, 161
Levi, Giuseppe, 265
Levi-Montalcini, Rita, 265, 266–267
Life Evolving (de Duve), 259
light, 167–171, 173
light therapy, 50
limited sloppiness, principle of, 192
linden flowers, 228
Lipkin, David, 217
Lipmann, Fritz Albert, 22, 26–28
liver, 47, 253–254
liver enzymes, 210
Loewi, Otto, 122, 208
Lorenz, Konrad, 227, 236–245
Louisiana Science Education Act, 260
lung cancer, 186
lungs, 47–48
Luria, Salvador, 187, 191, 195–198, 265, 267
Lwoff, André, 142–147, 151, 152
Lynen, Feodor, 133–137, 141
lysergic acid diethylamide (LSD), 57, 211–213
lysosomes, 256, 257
lysosomotropic agents, 257

Maass, Clara Louise, 6–7
Mach bands, 100–101
malaria, 4
Malaysia, 45
Malcolm Is a Little Unwell, 13
Mandela, Nelson, 121–122
Mandela Rhodes Foundation, 121–122
Manhattan Project, 106, 108, 109
Manson, Patrick, 4

Marburg virus, 154–155, 281

Marshall, Barry, 51

masked facies, 226

Massachusetts Institute of Technology (MIT), 174, 198, 277

mathematics, 170

Matilda effect, 81

Matthaei, J. H., 181

May, Alan Nunn, 108–109

MDCK cell line, 89–90

measles virus *(rubeola)*, 33–37, 39, 90

Mechnikoff, Elie, 50

meclizine, 257

medals, values of, 26, 118

Medawar, Peter, 87, 92, 95–96

Memphis, Tennessee, 2, 3

mengovirus, 277–278

meningitis, 15, 154

mental health, 214–215

The Merck Manual of Diagnosis and Therapy, 59

Merkel cell polyomavirus, 161

messenger RNA (mRNA), 79, 181–182, 263

metabolism
   of caffeine, 210
   cAMP and, 218
   discovery regarding, 22
   drug exposure and, 210–211
   enzymes and, 78–79
   insulin, 253–254
   process of, 24

metabolites, 209–210

Mexico, 2

Meyerhof, Otto, 171–172

miasma theory, 143

mice
   cancer research on, 156
   LSD research on, 212
   mengovirus and, 277

research regarding, 92, 242

viruses and, 90

yellow fever research on, 9

microbes, 14, 73, 151

Mill, John Stuart, 241–242

mitochondria, 251, 252, 263

Mittag-Leffler, Gösta, 42

Mobile, Alabama, 2

modulators, 167

molecules, 74–75

monkeys, 8, 9, 33, 212, 281, 282

Monod, Jacques, 142, 147–149, 151, 152

morphine, 210–211

mosquitoes, 3–4, 5–6, 7, 8

mumps, 31

Munich, Germany, 233

murder, 62–63

mushrooms, 133–134

mutual annihilation, 94

Nash, John, 26

nasopharyngeal carcinoma, 160

Nate D. Sanders, 25

National Cancer Institute, 274

National Institutes of Health (NIH), 116–117, 179, 182, 210, 211, 283

National Socialist German Lecturers League (NSDDB), 189

natural selection, 129–130, 132

nature, benefits of, 183

Nazis
   attacks by, 148–150, 206–208
   within Belgium, 254
   control by, 188–190
   doctors of, 240
   eugenics of, 239, 240

within France, 143–144

ideology of, 230–231

influence of, 51–52, 55

Konrad Lorenz and, 236–237

mass murders by, 53, 262, 266

media control by, 205

within the Netherlands, 246–247

rise of, 254

Nehru, Jawaharlal, 94

nerves/nervous system
   discoveries regarding, 122, 123
   language development and, 168
   light and, 167, 169–171
   nerve-conduction studies, 127–128
   norepinephrine within, 203–204
   research regarding, 120, 124–125, 126–127
   transmission of, 208–209

neural Darwinism, 225

neurophysiology, 127–128

New Orleans, Louisiana, 2

New York, 2

New Zealand, 45

Nicolle, Charles, 50

nicotine, 64–65

Niedergerke, Rolf, 129

Nirenberg, Marshall, 178–182

nitrogen, 105

Nitti, Filomena, 61, 65

Nitti, Francesco Saverio, 61

Nobel Prize
   1903 Nobel Prize, 50
   1908 Nobel Prize, 50
   1921 Nobel Prize, 218
   1923 Nobel Prize, 50
   1926 Nobel Prize, 26

1928 Nobel Prize, 50

1932 Nobel Prize, 120

1936 Nobel Prize, 122

1937 Nobel Prize, 171

1939 Nobel Prize, 51

1945 Nobel Prize, 15n2

1947 Nobel Prize, 216

1951 Nobel Prize, 1–13, 51

1952 Nobel Prize, 14–21

1953 Nobel Prize, 22

1954 Nobel Prize, 29–40

1955 Nobel Prize, 41–46

1956 Nobel Prize, 47–59

1957 Nobel Prize, 60–66

1958 Nobel Prize, 67–76

1959 Nobel Prize, 77–86

1960 Nobel Prize, 87–96

1961 Nobel Prize, 97–104

1962 Nobel Prize, 26, 105–119

1963 Nobel Prize, 26, 120–132

1964 Nobel Prize, 133

1965 Nobel Prize, 26, 142–152

1966 Nobel Prize, 153–165

1967 Nobel Prize, 166–177

1968 Nobel Prize, 178–186

1969 Nobel Prize, 187–201

1970 Nobel Prize, 202–215

1971 Nobel Prize, 26, 216–220

1972 Nobel Prize, 221–226

1973 Nobel Prize, 227–248

1974 Nobel Prize, 249

1975 Nobel Prize, 265–287

1994 Nobel Prize, 26

2005 Nobel Prize, 51

2006 Nobel Prize, 85–86

Noguchi, Hideyo, 8

non-Hodgkin's lymphoma, 161

noninnate behavior, 243

noradrenaline/norepinephrine, 203–204

Nosema disease, 232–233

nuclear weapons/nuclear war, 108–109, 176

Nuremberg Race Laws, 230

Ochoa, Severo, 77, 78–82, 181, 216

On Aggression, 243

"one gene, one enzyme" hypothesis, 70

Onsager, Lars, 186

organ transplant, 87

Orgel, Leslie, 113

origins of life, 258–259, 260

Orwell, George, 21

osteosarcoma, 165

O'Toole, Margot, 283–285

outdoor learning, benefits of, 183

Overton, Charles Ernest, 131

Oxford University, 24

painkillers, 209–210

Palade, George, 249, 261–264

panspermia, 113

panther cap mushrooms, 134

papain, 222

paralysis, 62

Pasteur, Louis, 3, 143

Pasteur Institute (Paris), 143, 145, 148, 149

Pasteur Institute of Tunis, 10

pasteurization, 143

Patent and Trademark Law Amendments Act (Bayh-Dole Act), 280

Paxil, 213

Peebles, Thomas, 33

penicillin, 15, 15n2

penis, 203

Pentz, Mike, 110

perceptual slovenliness, 238

Pétain, Philippe, 144

phages. *See* bacteriophages (phages)

pharmacogenomics, 210

Philadelphia, Pennsylvania, 7

phosphates, 105

phosphorylase, 79

Photo 51 (DNA), 107–108, 111–112

photoreactivation, 192

Phycomyces, 194

Planetary Protection Center of Excellence (Jet Propulsion Laboratories of the California Institute of Technology), 73–74

pneumococcus bacteria, 31

pneumonia, 15, 31

polarized light, 234–235

poliovirus, 31–33, 38–39, 90, 161, 278

Polish Selection, 241

polymerase enzyme, 86

polyphosphate kinase (PPK) enzyme, 85

Popper, Karl, 125, 126

popularity, medical science and, 58

Porter, Rodney, 221–223

posttransplant lymphoproliferative disorder, 161

potential pandemic pathogens (PPPs), 89, 90–91

poverty, 130–131

Press, Elizabeth (Betty), 222–223

priesthood, medical, 58

principle of limited sloppiness, 192

prions, 155

prophages, 145–146

prospective concept, 56

prostaglandin, 203

prostate cancer, 163

Protection of Science and Learning (SPSL), 23–24, 25

proteins, 22, 151, 178–186, 200

protons, 68

provirus, 273

Prozac, 213

Psychologists for Peace, 110

purging, 153

pyrazinamide, 21

Quaker Oats, 34

quinine, 257

rabbits, 157

rabies, 143

Ramón y Cajal, Santiago, 122

rats, 254

Rats, Lice and Men (Zinsser), 30

rebellion, 243

Reed, Walter, 4, 5, 7–8

regulatory gene, 150–151

Rehn, E., 22

research, regulations regarding, 283

researchers

controversies regarding, 283–285

deaths of, 1, 4–5, 6–7

See also specific people

retina, 171–172, 173

retroviruses, 275

reverse transcriptase, 274

Reviewing Committee on the Export of Works of Art and Objects of Cultural Interest (RCEWA), 25

Reye's syndrome, 40

Rhodes, Cecil John, 121

Rhodes Scholarship, 121–122

rhodopsin, 171, 173

ribonucleic acid (RNA)

DNA and, 86, 178, 274, 279

enzymes, 278–279

HIV and, 275

messenger RNA (mRNA), 79, 181–182, 263

replicase, 279

research regarding, 77, 78–80, 178

Rous sarcoma virus (RSV) and, 273

transfer RNA (tRNA), 178, 185, 263

viruses and, 84, 277–278

ribosomes, 263

Richards, Dickinson, 47, 55–56, 57–59

Rickettsia bacteria, 31

rifampin, 21

RNA polymerase, 79

RNA Tie Club, 180–181

Robbins, Frederick, 29, 32, 39–40

robinia flowers, 228

Rockefeller Foundation, 11, 23

Rockefeller Institute, 8

Röhm Purge, 52

Romania, 262

Roosevelt, Franklin D., 31

Rose, Steven, 110

Ross, Ronald, 4

Rossiter, Margaret, 81

rotavirus vaccine, 90

Rous, Peyton, 153–160, 251

Rous sarcoma virus (RSV), 157–158, 160, 272–273

rubeola (measles virus), 33–36

Rusk, Dean, 175

Russell, Bertrand, 110

Russell-Einstein Manifesto, 110

Rust, Bernhard, 189

Rutgers Research and Endowment Foundation, 19

sabbaticals, 185

Sabin, Albert, 32

Sakharov, Andrei, 176

Salk, Jonas, 32, 38

Salk, Peter, 32

Salk Cancer Center, 185–186

Salk Institute, 270, 279

Sanofi Pasteur, 13

sarcoma virus, 250–252, 268

Savannah, Georgia, 2

Schatz, Albert, 14, 16–20, 21

Schneider, Dr., 49

Schoenheimer, Rudolf, 139

Schutzstaffel (SS), 240

Schweitzer, Albert, 10

Scientists Against Nuclear Arms (SANA), 110

Scientists for Global Responsibility (SGR), 110

second messenger effect, 217

selective serotonin reuptake inhibitors (SSRIs), 213, 214–215

The Self and Its Brain (Eccles and Popper), 125

self-conscious mind, 125

self-experimentation, 47, 48, 50–51

sensations, 100–101, 103–104

serotonin, 211, 213–215

17DD vaccine, 12–13

sex hormones, 165

sexuality, 116

sheep, 155

Sherrington, Charles, 120, 166, 208

shingles *(zoster)*, 39, 88

Shope, Richard, 157

silkworm disease, 143

Simian virus 40 (SV40), 161

Singapore, 45

singularities, 259, 260–261

Sir Hans Krebs Trust, 25

skin color, 116

skin transplants, 92

sliding muscle fiber theory, 131

smallpox *(Vaccinia)* virus, 31

smelling, 103, 228–229

Smith, Kline & French, 255

smoking, 64–65, 94, 275

social activism, 110

soil bacteria, 15

soil conservation, 14

Sommerlath, Walther, 205

sound, 97–104

space, biochemical analysis of, 73

Spanish-American War, 4

Spanish flu epidemic, 89

special relativity, 218

spontaneous generation, theory of, 143

squid, 126–127

squiggle symbol, 27

Starr, Ringo, 21

statin, 140–141

sterilizations, 52, 282–283

stimulants, 210

stomach ulcers, 51

Stravinsky, Igor, 16

Streptomyces griseus, 18

streptomycin, 14, 18–20, 21

Strychnos toxifera, 61

S type personality, 238

substance P, 203

succinylcholine (sux), 61–63

sugars, 105, 147–148

sulfa antibiotic precursor, 50–51

sun/sunlight, 183, 234–235

Sutherland, Earl Wilbur, Jr., 216–220, 256

SV40 animal virus, 281

Sweden, 43–45, 166, 205

T1 bacteriophage, 193

tasting, 103

Tatum, Edward Lawrie, 69–70, 75

T cell lymphomas, 161

telomeres, 270

Temin, Howard, 265, 268, 272–276, 279–280

Temin Effect, 274

testosterone, 163

Theiler, Max, 1, 8, 9–11, 51

Theorell, Hugo, 41–46, 255

theory of spontaneous generation, 143

Thorazine (chlorpromazine), 57

three-spined stickleback, 247

thyroid cancer, 165

thyroid gland, 256

Tinbergen, Nikolaas, 227, 236, 245–248

tobacco, 64–65, 94

tooth decay, 43

transduction, 72

transfer RNA (tRNA), 178, 185, 263

transfusions, blood, 157

Treaty of Versailles, 138

Trinity College, 128

tuberculosis (TB)
    AIDS and, 16
    corpses of, 154
    deaths from, 21
    defined, 153
    gram-negative bacteria and, 15
    mortality rate of, 16
    overview of, 15–16
    polyphosphate kinase (PPK) enzyme and, 85
    streptomycin as treatment for, 14
    symptoms of, 153
    treatments for, 21
    tumors, 155–156, 157, 158, 164, 165, 238, 251, 265–287. *See also* cancer

Turner, J. R., 157

typhoid, 154

typhus, 30, 50

Ullmann, Agnes, 149

ultraviolet light, 192

United States, 2, 4, 45

University of Sheffield, 24

Usmanov, Alisher, 26

US military, 11

vaccination, 37, 90–91

vaccines
    a-cellular pertussis, 37
    AIDS, 275

avian leukosis virus and, 11–12

contamination of, 161

COVID-19, 38

Edmonston-Moraten vaccine strain, 35

efficacy *versus* effectiveness, 36

flu, 36, 91–92

French vaccine, 10

measles, 33–37, 90

polio, 32–33, 38–39, 90, 161

rabies, 143

Rockefeller Foundation and, 11

rotavirus, 90

self-experimentation regarding, 51

17DD, 12–13

typhus, 30

yellow fever, 8, 9–11, 12–13, 51, 162

Vaccinia (smallpox) virus, 31

van Parijs, Luk, 286

varicella (chicken pox), 31, 39, 88

Venetian paintings, 139

Vietnam War, 174–176

viral infection, 146

virulent bacteriophages, 145–146

viruses

adenoviruses, 281

AIDS, 16, 275

within animals, 277–278, 281

antibodies within, 221

attenuation and, 9

avian leucosis virus (ALV), 11–12, 162

bacteria and, 142, 145, 190

bird flu virus, 90–91

cancer associations of, 160–162, 273–274

cell/virus system, 146

chicken pox *(varicella)*, 31, 39, 88

composition of, 84

COVID-19, 2, 38, 155, 257, 286

coxsackie infection, 39

Coxsakieviruses, 39

Creutzfeldt-Jakob disease (CJD), 155

DNA and, 84, 268

DNA virus, 84

Ebola-Reston, 282

Ebola virus, 154–155, 282

Epstein-Barr Virus (EBV), 160–161

escape of, 88–91

etymology of, 146–147

filterable particles, 147, 156, 160

formation of, 68

H1N1, 89

H5N1, 90–91

human immunodeficiency virus (HIV), 154, 161, 275

human papillomavirus (HPV), 161

Human T-cell lymphotrophic virus type 1 (HTLV-1), 161

influenza, 36, 91–92

influenza A, 31

intubation of, 89–90

Kaposi's sarcoma-associated herpesvirus, 161

lab escape of, 281–283

manipulation of, 88–89

Marburg virus, 154–155, 281

measles virus *(rubeola)*, 33–37, 39, 90

mengovirus, 277–278

Merkel cell polyomavirus, 161

mosquito transmission of, 3–4, 5–6, 7, 8

mumps, 31

poliovirus, 31–33, 38–39, 90, 161, 278

prions, 155

prophages, 145–146

provirus, 273

quantification of, 272

retroviruses, 275

RNA and, 84, 277–278

Rous sarcoma virus (RSV), 157–158, 160, 272–273

rubeola (measles virus), 33–36

sarcoma, 250–252

sarcoma virus, 250–252, 268

shingles *(zoster)*, 39, 88

Simian virus 40 (SV40), 161

smallpox *(Vaccinia)* virus, 31

Spanish flu epidemic, 89

SV40 animal virus, 281

T1 bacteriophage, 193

within test tubes, 84

vaccination introduction of, 90–91

Vaccinia (smallpox) virus, 31

varicella (chicken pox), 31, 39, 88

viral infection, 146

virulent bacteriophages, 145–146

zoster (shingles), 39, 88

See also bacteriophages (phages); specific viruses

vision, 103–104, 167–172

vitamin A, 171, 173

vitamin B9 (folate), 83

vitamin B12, 47

vitamins, 78, 151

Vogt, Marguerite, 267, 269–270

voice production, 168

von Békésy, Georg, 97–104

von Euler, Ulf, 202–206

von Euler-Chelpin, Hans, 202

von Frisch, Karl, 227–235

Wahoo, Nebraska, 67–68

Waksman, Selman, 14, 17–18, 19–20

Wald, Elijah, 173–174, 177

Wald, George, 166, 171–177

Wallace, Henry, 197

Wallgren, A., 21

Warburg, Otto, 42

water, flouridation of, 43–45

Water Flouridation Act, 45

Watson, James, 26, 95, 105, 111, 114–119, 197, 283

Watson, Rufus, 118–119

Wellbutrin, 213

Weller, Thomas, 29, 31, 32, 38–39

Wertham, Fredric, 245

Wilkins, Maurice, 105, 106–109

Wilson, E. O., 114

Wolfe, Thomas, 16

Wollman, Elie, 144, 145, 150

Wollman, Eugene and Elisabeth, 142, 144, 196

women, higher education employment of, 72

Wood, Leonard, 5

World War I, 187, 249–250

World War II
Bernard Katz within, 206–208
within France, 143–144, 148–150, 195–196

within Italy, 266

Karl von Frisch within, 230–231

within Louvain, 254

Max Delbrück within, 188–191

within Munich, Germany, 233

within the Netherlands, 246–247

within Russia, 266

Soviet Union within, 108–109

Sweden within, 205

Wuhan, China, 282

X-rays, 48, 70

yeast, 133, 134

yellow fever
animal research regarding, 8, 9
avian viruses and, 12
COVID-19 compared to, 2
discoveries regarding, 1
human experimentations regarding, 4–7
informed consent regarding, 4–5
mosquitoes and, 3–4, 5–6, 8
spiraled bacterium of, 8
spread of, 2
symptoms of, 1–2
vaccine for, 8, 9–11, 12–13, 51, 162

youth rebellion, 243

Yugoslavia, 281

Zamenhof, L. L., 60

Zanuck, Daryl, 67

Zimmer Lederberg, Esther, 71

Zinder, Norton, 72

Zinsser, Hans, 30

Zoloft, 213

zoster (shingles), 39, 88

ABOUT THE AUTHOR

Moira Dolan, MD, is a graduate of the University of Illinois School of Medicine and has been a practicing physician for over 30 years. Dr. Dolan is a patient advocate and public speaker who educates patients on their rights and the need for a healthy skepticism of the medical profession. In addition to being the author of *Boneheads and Brainiacs: Heroes and Scoundrels of the First 50 Years of the Nobel Prize in Medicine* (volume 1 of the *Boneheads and Brainiacs* series), she is the author of *No-Nonsense Guide to Antibiotics, Dangers, Benefits & Proper Use*; *No-Nonsense Guide to Cholesterol Medications, Informed Consent and Statin Drugs*; and *No-Nonsense Guide to Psychiatric Drugs, Including Mental Effects of Common Non-Psych Medications*. In addition, Dr. Dolan is a contributor to the blog SmartMEDinfo. She maintains a private medical practice in Austin, Texas.

Enjoy more great medical biographies by reading *Boneheads and Brainiacs: Heroes and Scoundrels of the Nobel Prize in Medicine*, book one in Moira Dolan's series about the triumphs and follies of Nobel Prize in Medicine winners.

Price: $18.95 US
ISBN: 978-1-61035-350-2
Available in paperback and ebook editions.